THE HISTORY of MEDICINE

THE MIDDLE AGES

500–1450

THE HISTORY of MEDICINE

THE MIDDLE AGES

500–1450

KATE KELLY

THE MIDDLE AGES: 500–1450

Facts On File, Inc.
An imprint of Infobase Publishing
132 West 31st Street
New York NY 10001

Library of Congress Cataloging-in-Publication Data
Kelly, Kate, 1958–
The Middle Ages: 500–1450 / Kate Kelly.
p. cm.—(The history of medicine)
Includes bibliographical references and index.
ISBN-13: 978-0-8160-7206-4
ISBN-10: 0-8160-7206-X
1. Medicine, Medieval. I. Title.
R141.K45 2010
610.94'0902—dc22 2008048709

Facts On File books are available at special discounts when purchased in bulk quantities for businesses, associations, institutions, or sales promotions. Please call our Special Sales Department in New York at (212) 967-8800 or (800) 322-8755.

You can find Facts On File on the World Wide Web at http://www.factsonfile.com

Text design by Annie O'Donnell
Illustrations by Bobbi McCutcheon
Photo research by Elizabeth H. Oakes

Printed in the United States of America

Bang Hermitage 10 9 8 7 6 5 4 3 2 1

This book is printed on acid-free paper.

CONTENTS

PREFACE

"You have to know the past to understand the present."
—*American scientist Carl Sagan (1934–96)*

The history of medicine offers a fascinating lens through which to view humankind. Maintaining good health, overcoming disease, and caring for wounds and broken bones was as important to primitive people as it is to us today, and every civilization participated in efforts to keep its population healthy. As scientists continue to study the past, they are finding more and more information about how early civilizations coped with health problems, and they are gaining greater understanding of how health practitioners in earlier times made their discoveries. This information contributes to our understanding today of the science of medicine and healing.

In many ways, medicine is a very young science. Until the mid-19th century, no one knew of the existence of germs, so as a result, any solutions that healers might have tried could not address the root cause of many illnesses. Yet for several thousand years, medicine has been practiced, often quite successfully. While progress in any field is never linear (very early, nothing was written down; later, it may have been written down, but there was little intra-community communication), readers will see that some civilizations made great advances in certain health-related areas only to see the knowledge forgotten or ignored after the civilization faded. Two early examples of this are Hippocrates' patient-centered healing philosophy and the amazing contributions of the Romans to public health through water-delivery and waste-removal systems. This knowledge was lost and had to be regained later.

The six volumes in the History of Medicine set are written to stand alone, but combined, the set presents the entire sweep of the history of medicine. It is written to put into perspective for high school students and the general public how and when various medical discoveries were made and how that information affected health care of the time period. The set starts with primitive humans and concludes with a final volume that presents readers with the very vital information they will need as they must answer society's questions of the future about everything from understanding one's personal risk of certain diseases to the ethics of organ transplants and the increasingly complex questions about preservation of life.

Each volume is interdisciplinary, blending discussions of the history, biology, chemistry, medicine and economic issues and public policy that are associated with each topic. *Early Civilizations,* the first volume, presents new research about very old cultures because modern technology has yielded new information on the study of ancient civilizations. The healing practices of primitive humans and of the ancient civilizations in India and China are outlined, and this volume describes the many contributions of the Greeks and Romans, including Hippocrates' patient-centric approach to illness and how the Romans improved public health.

The Middle Ages addresses the religious influence on the practice of medicine and the eventual growth of universities that provided a medical education. During the Middle Ages, sanitation became a major issue, and necessity eventually drove improvements to public health. Women also made contributions to the medical field during this time. The Middle Ages describes the manner in which medieval society coped with the Black Death (bubonic plague) and leprosy, as illustrative of the medical thinking of this era. The volume concludes with information on the golden age of Islamic medicine, during which considerable medical progress was made.

The Scientific Revolution and Medicine describes how disease flourished because of an increase in population, and the book describes the numerous discoveries that were an important aspect of this time. The volume explains the progress made by Andreas Vesalius (1514–64) who transformed Western concepts of the structure of the human body; William Harvey (1578–1657), who studied and wrote about the circulation of the human blood; and Ambroise Paré (1510–90), who was a leader in surgery. Syphilis was a major scourge of this time, and the way that society coped with what seemed to be a new illness is explained. Not all beliefs of this time were progressive, and the occult sciences of astrology and alchemy were an important influence in medicine, despite scientific advances.

Old World and New describes what was happening in the colonies as America was being settled and examines the illnesses that beset them and the way in which they were treated. However, before leaving the Old World, there are several important figures who will be introduced: Thomas Sydenham (1624–89) who was known as the English Hippocrates, Herman Boerhaave (1668–1738) who revitalized the teaching of clinical medicine, and Johann Peter Frank (1745–1821) who was an early proponent of the public health movement.

Medicine Becomes a Science begins during the era in which scientists discovered that bacteria was the cause of illness. Until 150 years ago, scientists had no idea why people became ill. This volume describes the evolution of "germ theory" and describes advances that followed quickly after bacteria was identified, including vaccinations, antibiotics, and an understanding of the importance of cleanliness. Evidence-based medicine is introduced as are medical discoveries from the battlefield.

Medicine Today examines the current state of medicine and reflects how DNA, genetic testing, nanotechnology, and stem cell research all hold the promise of enormous developments within the course of the next few years. It provides a framework for teachers and students to understand better the news stories that are sure to be written on these various topics: What are stem cells, and why is investigating them so important to scientists? And what is nanotechnology? Should genetic testing be permitted? Each of the issues discussed are placed in context of the ethical issues surrounding it.

Each volume within the History of Medicine set includes an index, a chronology of notable events, a glossary of significant terms and concepts, a helpful list of Internet resources, and an array of historical and current print sources for further research. Photographs, tables, and line art accompany the text.

I am a science and medical writer with the good fortune to be assigned this set. For a number of years I have written books in collaboration with physicians who wanted to share their medical knowledge with laypeople, and this has provided an excellent background in understanding the science and medicine of good health. In addition, I am a frequent guest at middle and high schools and at public libraries addressing audiences on the history of U.S. presidential election days, and this regular experience with students keeps me fresh when it comes to understanding how best to convey information to these audiences.

What is happening in the world of medicine and health technology today may affect the career choices of many, and it will affect the health care of all, so the topics are of vital importance. In addition, the public health policies under consideration (what medicines to develop, whether to permit stem cell research, what health records to put online, and how and when to use what types of technology, etc.) will have a big impact on all people in the future. These subjects are in the news daily, and students who can turn to authoritative science volumes on the topic will be better prepared to understand the story behind the news.

ACKNOWLEDGMENTS

This book as well as the others in the set was made possible because of the guidance, inspiration, and advice offered by many generous individuals who have helped me better understand science and medicine and their histories. I would like to express my heartfelt appreciation to Frank Darmstadt, whose vision and enthusiastic encouragement, patience, and support helped shape the series and saw it through to completion. Thank you, too, to the Facts On File staff members who worked on this set.

The line art and the photographs for the entire set were provided by two very helpful professionals—artist Bobbi McCutcheon provided all the line art; she frequently reached out to me from her office in Juneau, Alaska, to offer very welcome advice and support as we worked through the complexities of the renderings. A very warm thank you to Elizabeth Oakes for finding a wealth of wonderful photographs that helped bring the information to life. Carol Sailors got me off to a great start, and Carole Johnson kept me sane by providing able help on the back matter of all the books. Agent Bob Diforio has remained steadfast in his shepherding of the work.

I also want to acknowledge the wonderful archive collections that have provided information for the book. Without places like the Sophia Smith Collection at the Smith College Library, firsthand accounts of the Civil War battlefield treatment or reports such as Lillian Gilbreth's on helping the disabled after World War I would be lost to history.

INTRODUCTION

"No lepers, lunatics or person's having the falling sickness or other contagious disease, and no pregnant women, or sucking infants and no intolerable persons, even though they be poor and infirm, are to be admitted."

—Stated Rules of St. John's Hospital, Bridgwater, England, 1210

The fall of the Roman Empire set off a chain of events that affected all aspects of progress during the Middle Ages, including medical progress. In the West, any sort of patientcentric care that had developed during earlier times was swallowed up by folk medicine that featured astrological analysis and experimentation with various plants and herbs. Added to these approaches were healing prayers, magic spells, and various forms of mysticism.

The Eastern Empire split from the Roman Empire in 286 C.E., so development in the East occurred independently from what was happening in the West. The Western Empire was mired in economic and political difficulties, but the Byzantine Empire developed into a successful civilization that maintained an organized practice of medicine. Had Galen and Hippocrates not been adapted and translated by Arab scholars for use in the East, their teachings might have disappeared. Later on, the inhabitants of the West began to appreciate that the Islamic people had a better understanding of health and wellness, and eventually these medical philosophies spread back to the West where they had originated.

The Middle Ages: 500–1450 focuses on medical developments during the time period of the years 500–1450. During the early Middle Ages (529–800) and the middle time period (ca. 800–1100), the rise of Christianity had a definite effect on the practice of medicine. Pope Gregory (ca. 540–604) stressed that prayer was more important than medicine, and over time, that sentiment became pervasive. Every time a person got better, it was regarded as a miracle. As different schools of thought developed, tension arose between church-related cures and folk medicine. The church taught that since God sometimes sent illness as punishment, then prayer and repentance could lead to recovery. When Christians used herbal remedies, the church wanted the magic spells to be replaced with prayers of devotion.

During the High Middle Ages (ca. 1200s–1400), the West went through a period of economic and political renewal that eventually led to a shift in the practice of medicine. Agriculture methods improved so fewer people were needed to raise food to feed a population, causing more people to move to towns and cities and giving them time for education. Eventually university training became valued, and the medical theories and ideas preserved by the scholars from the Byzantine Empire became the basis for university education in the medical field.

Chapter 1 presents the basic perceptions of medicine during this time: how people thought disease was transmitted, how religion affected medical progress, and the importance of a proper garden for growing both food to eat and substances to be used in medicine.

Chapter 2 focuses on the various categories of medieval healers—from folk healers to those who were trained at a university. Chapter 3 talks about the different diagnostic methods used during medieval times and also explains the range of treatments, from religious healing to herbal medicines. Herbal cures were found through trial and error, and medieval healers often started by pairing "like with like" to identify usefulness. For example, the seeds of skullcap can look like small skulls, so skullcap was used as a headache remedy.

Surgical knowledge of the day is explored in chapter 4. Surgery was still viewed as a tradesmanlike activity and therefore, most physician-healers refused to undertake it. (Surgery was also high risk. Since healers' fees were based on their reputations, there was good reason not to practice an aspect of medicine with a high failure rate.) Town barbers learned to manage bleeding because of shaving-related cuts and nicks, and this knowledge brought people with injuries to barbers for help and advice. Eventually barber-surgeons became an acknowledged specialty.

Women were important in the practice of medieval medicine. They were the "constants" of the community, and their knowledge grew with experience. Chapter 5 outlines women's contributions to medicine and some of the prominent women practitioners are profiled for their accomplishments.

Chapter 6 addresses the early public health laws and practices. From ways to bring in clean water to methods for getting rid of waste, the medieval people made progress in some areas but took steps back in others.

When the Black Death occurred in Europe in the 14th century, Europeans were still relying on religious theories, and as a result about one-third of the population of Europe died from the plague. Chapter 7 examines how this devastating disease was handled by medieval communities and practitioners and why it was so lethal. Smallpox and leprosy were also major issues of the time, and these diseases are also discussed.

Chapter 8 examines the contributions of the Byzantine Empire to the future of medicine, and introduces Ibn Sina (Avicenna), a physician whose contributions rank with Galen.

Many of the healing methods used during the Middle Ages are very primitive by today's standards, and the people of that day still had no understanding of the role that germs played in disease. While there is no doubt that the heavy emphasis on prayer and pilgrimages as possible cures slowed the progress of medicine, medieval people still had to overcome illnesses and injuries, and in the process, discoveries were made. Some of the herbal cures used have stood the test of time, and while the surgical methods were primitive, the learning that was achieved through what they undertook opened the pathway for later developments in the field.

The Middle Ages: 500–1450 illuminates what occurred during medieval times that affected future developments in medicine. The back matter of the volume contains a chronology, a glossary, and an array of historical and current sources for further research. These sections should prove especially helpful for readers who need additional information on specific terms, topics, and developments in medical science.

This book is a vital addition to literature on the Middle Ages because it provides readers with a better understanding of the accomplishments of the time and explains how and why scientific understanding was poised for the breakthrough of the Renaissance period.

1

Medical Beliefs in Medieval Times

The fall of the Western Roman Empire altered the pace of progress for the people of the medieval era. Parts of Europe were so disrupted that chaos overshadowed any type of scientific or medical knowledge. In other areas, the culture held together well enough that some of the medical practices from Greek and Roman times survived to be passed on to later generations.

Inheriting these classical beliefs was a mixed blessing. The theories of *anatomy,* circulation, and how diseases passed among people were a good start at medical understanding, and the ideas would have been good "stepping stones" on the way to more advanced scholarship. Unfortunately, most practitioners viewed the information as indisputable truths, so the classic beliefs smothered the possibility of additional learning. Particularly damaging was the strict adherence to the belief in the four *humors,* and the resulting practice of *bloodletting* as a way to bring the humors in better balance.

Religious healing predominated during this time, and advances were made in botany, some of which led to the creation of herbal medications that were helpful. While *surgery* was held in low regard, it was vital when someone suffered *kidney stones* or was injured in a battle. The necessity of experimenting with various surgical remedies eventually led to some progress in this area.

This chapter focuses on what was happening in Western Europe during the early Middle Ages and outlines what medical practices were important during this time. The theories of *contagion* and the resulting decisions as to what would be curative are introduced. The importance of gardening to people of all classes is explained; religion's influence on medicine, and the subjects of *astrology* and *alchemy*—two of the fields of "science" that were pursued at this time—also are examined.

BELIEF IN THE FOUR HUMORS CONTINUES

The four humors and the importance of keeping them in balance in order to maintain good health was the most influential of the theories that were passed down from the Greeks and Romans; it was actually a dominant factor in medicine until the 19th century. A *physician* or healer was to assess the basic composition of each sick person and then establish a method to rebalance the patient's humors for better health. The factors that could affect humoral balance ranged from diet and the environment to the position of the stars.

The humors that constituted this theory were blood, *phlegm,* yellow *bile,* and black bile, and each had a very specific influence on the body noted as follows:

Blood: The qualities of this humor were hot and wet, and people who fell within this category were thought to be sanguine (in Latin *sanguis* means "blood") or have hopeful personalities. People who were thought to be dominated by this humor had ruddy, healthy complexions and were considered cheerful, warm, and generous. Medieval books usually portrayed a nobleman in this category because the ideal nobleman was expected to possess these qualities.

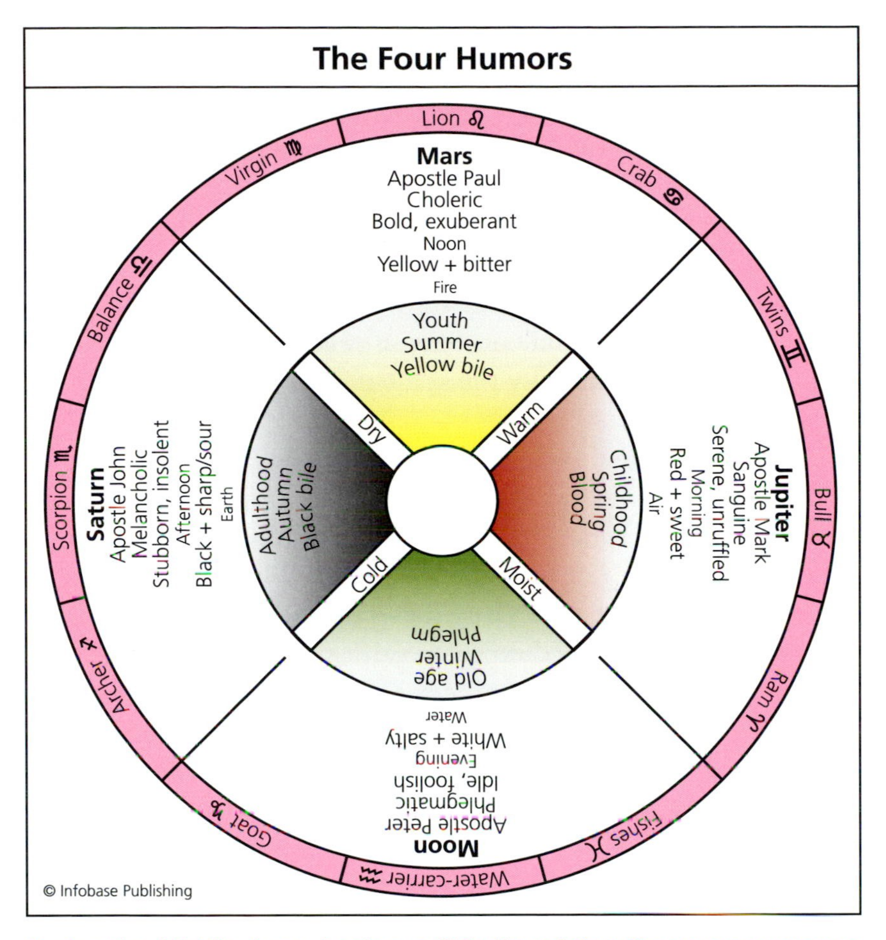

During the Middle Ages, healers still believed that the way to restore a person's health was to rebalance their humors.

Phlegm: The qualities of a person who was categorized as being dominated by phlegm were cold, wet, and slow to anger but also sluggish and dull. Generally speaking, phlegmatic people were considered detached and cool, and in medieval times it was the merchant who best fit this category.

Yellow bile: The qualities related to yellow bile were hot and dry. An excess of yellow bile (choler) was thought to make a person irritable. Thin people were categorized as choleric because of being in a state of constant agitation. Knights were frequently the symbols of those with yellow bile.

Black bile: Cold and dry were the elements related to black bile. The person with too much black bile was gloomy, depressed, and melancholy (this word is derived from the Greek words *melanos* meaning "black," and *chole* meaning "bile"). The person with black bile was usually cowardly, pale, and covetous of others. The pasty-faced scholar was generally the symbol of this personality.

Galen's analytical system for assessing and balancing the four humors became a lasting part of medical teachings: For example, too much phlegm caused lung problems, so physicians needed to find ways to rid the body of phlegm. These adjustments were usually achieved through diet, medicines, and bloodletting. Galen also taught that each humor connected with two of the four primary qualities of hot, cold, wet, and dry. To restore the balance of the humors, he recommended a system of opposites. Fever was treated with something cold, weak people were given difficult exercises to build up strength, and those with chest weakness were told to perform singing exercises, for example.

Galen differed from Hippocratic practice in one aspect of treatment: While Hippocrates felt that balance needed to be achieved in the body overall, Galen introduced the concept that balance could be reached organ by organ. This allowed physicians to develop organ-specific remedies, and this *philosophy* had a powerful influence on medicine.

To devise ways to maintain humoral balance, a physician or healer needed to maintain an almost encyclopedic understanding of how various aspects of life were categorized. Food, weather conditions, the seasons, and what people wore were all thought to affect humoral balance. For example, wool clothing was identified as warm and dry; winds from the north were classified as quite cold and somewhat dry. Spring was viewed as moist, and winter was cold.

While much of this belief offered very little in the way of the power to heal, most of it was harmless, and occasionally it was helpful because the method encouraged preventive health care. The followers of the humoral theory did not like waiting for a person to get sick and then deal with it. They viewed good health as an ongoing effort that needed to be tended to regularly. Medical advice usually included recommendations on diet, bathing (in hot, warm, or cold water, depending on the physician's *diagnosis* of the patient's humoral balance), proper environment, rest, and exercise. The dietary recommendations may have promoted better health simply by introducing greater variety into the diet, and being reminded to wear warm and dry, heavy clothing when the north wind blew was not a bad thing. Though medieval people were seriously misguided about much of their knowledge concerning the workings of the body and the way the human body fights off disease, some of the early efforts at maintaining good health were steps in the right direction.

Galen's beliefs influenced medicine throughout the Middle Ages *(National Library of Medicine)*

The humoral theory included the use of certain medications, and these were usually compounded from vegetable matter, most commonly from herbs. (For more information on this, see chapter 3, "Diagnosis and Treatment.")

Other Basic Beliefs

From Galen, medieval healers acquired a misguided understanding of the function of the blood. Galen believed that blood carried pneuma, or life spirit, and that the circulatory system permitted blood to pass through a porous wall between the ventricles of the heart. His teachings delayed the understanding of both circulation and physiology. He also caused a misunderstanding of anatomy; his knowledge was incorrect since he assumed that *dissecting* an ape provided information that directly related to human anatomy.

In contrast, Galen's experimentation led to knowledge of the workings of the nerves of the spinal cord, giving very valuable information that was carried on. He personally had excelled at diagnosis and prognosis of an illness, and some medieval physicians also acquired an ability to analyze the severity of an illness.

OVERALL HEALTH STATUS OF MEDIEVAL PEOPLE

The overall health of people who lived in the Middle Ages was not particularly good; people were expected to live about half as long as they do today. Urban dwellers in Italy in 1480 had a life expectancy of only about 30 years. People who lived in the countryside might live a little longer, perhaps to the age of 40.

People did not die of degenerative illnesses such as cancer as they do now; they died of illnesses like diarrhea, *smallpox, tuberculosis,* and *measles,* or they died from accidents, drowning, or being burned in a fire. Other illnesses that were common included *leprosy, plague,* typhoid, problems with parasites, recurring fevers of all types, "stone" (kidney stones), *dysentery,* and venereal disease.

Women lived shorter lives than men. Childbirth was risky, and women also suffered health problems from iron deficiency since there was little animal protein in the diet and women lost a lot of iron during menstruation. Not until late in the 15th century did women begin to outnumber men.

HOW MEDIEVAL PRACTITIONERS THOUGHT DISEASE WAS TRANSMITTED

During the Middle Ages there were many theories on what caused disease to spread. Some thought ailments were caused by the wrath of God; others thought witches and demons sent illnesses. Many believed in a strong astrological influence on life, and they felt that the planets and stars sometimes aligned to create bad health among people on earth. A few realized that swamps, filthy water, and areas where animal carcasses were left to decay could spread disease, but they didn't understand how these conditions gave rise to illness. They thought this environment led to spontaneous generation of some diseases by *miasma* (a cloud of noxious vapors emitted by the earth or formed in swamps or other areas of decay). This contaminated the air and water and could make people ill.

Attributing illness to a specific cause gave people a feeling of control over their lives, and if no other explanation presented itself, scapegoats often received the blame. Lepers and other "evil people" were thought to poison wells or take other actions that spread disease, and of course, if a scapegoat was identified, the townspeople made that person's life miserable.

The discovery of *microbes* did not occur until Antoni van Leeuwenhoek (1632–1723), a Dutch cloth merchant whose hobby involved handcrafting microscopes, eventually saw "little animacules" through the magnifying lenses he created. Even then, scientists of that day had no understanding about the role microbes played in disease. Bacteria and its part in spreading illness were not even suspected until the 18th century, and it took until the 19th century for the theory to be accepted.

Despite not understanding how illness spread, medieval people realized that diseases could be contagious, and this led to some customs that helped reduce the spread of illness. Inhabitants of port towns began to note that when ships from out of the area arrived with crewmen who were sick, sometimes these men came on shore and townspeople soon suffered similar symptoms. As a result, port cities developed a rule that required ships to anchor offshore for up to 40 days before the crew could come on land. This method was first used in the ports of Italy and southern France in the late 14th century as an effort to combat recurrences of the *bubonic plague* (the Black Death), and the word *quarantine* is derived from the Italian word for a period of 40 days. Later on, quarantines were used to require sick people to stay in their homes for a period until they recovered or died. (As recently as the early 20th century, towns in the United States still quarantined people in their homes for illnesses such as scarlet fever.)

TYPES OF CURES USED

With a poor understanding of what caused disease, the medieval practitioner's theories on what would cure illnesses was understandably misguided. Diseases brought about by the wrath of god should be cured through prayer or a *pilgrimage*; witches and demon-caused illnesses needed magic potions, charms, and the casting of spells. Diseases spread by miasma (noxious vapors) might be cured or prevented by burning herbs and incense. Other healing methods employed regularly during this period included the following:

- Bleeding and *cupping.* These processes were performed curatively and also used preventively. Medieval practitioners seemed to recommend bloodletting be done every three months or so. Cupping involved using a hot cup to draw blood and other fluids out of the body or away from a certain body part.
- *Cautery.* Cautery burns the body and was used to clean and seal a cut.
- Purgatives. Laxatives and emetics were used to clean out the organs or to bring about balance of the humors.
- Vermifuges. Intestinal worms were a big problem during this time, and treatment involved the use of herbal remedies with nerve-paralyzing properties so that the worms would detach. This treatment would be used in combination with

Physicians often recommended that people leave town and move to the countryside for a time if signs of sickness appeared among townspeople. Those who could not leave town—including

laxatives so that the body could be rid of the parasites.

- Baths, steam baths, and fumigations. Some believed that bathing and water were bad for one's health because it opened the *pores,* which could cause illness, but many others believed that bathing was healing.
- "Simples" and compound remedies. Simples were one-ingredient medicines, and these were used as frequently as compound remedies. Medicines themselves were mixed in to be used in many forms that could be consumed as syrups, electuaries, waters, wines, or used as topical ointments or plasters.
- Surgery. Because certain conditions required a surgical solution, surgeries were performed. The outcome was frequently poor, so physicians left this type of work to others, generally barber-surgeons.
- Dietary changes. These were considered curative as part of balancing the humors.
- Charms and relics. These were used at times when an illness was thought to have been caused by some type of bad spirit.

Trial and error were at the heart of all the methods used. If a patient improved, then it was an excellent remedy and would be tried on others; if the patient did not get better, then the healer had to evaluate whether it was the cure that failed, or whether the patient was "at fault" in some way.

the physicians who stayed to treat the sick—wore pomanders that contained aromatic compounds, or they burned incense; both were thought to neutralize the air and prevent illness. Physicians also wore long robes and donned masks with large birdlike beaks filled with aromatic spices in order to maintain their health. Though this did little to prevent the spread of illnesses, this practice persisted into the 18th century.

THE INFLUENCE OF RELIGION ON MEDICINE

Perhaps because the medical treatments that were attempted in the early Middle Ages were so ineffective, medieval people began to look for new answers. Over time, many healers concluded that disease was destiny or was punishment for a transgression of the patient's. For this reason, spiritual intervention was often considered more important than any physical remedies. While the church believed that the sick may have needed to be made comfortable by a healer, the church elders generally felt that being blessed by the priest was likely to be more effective at healing. In 1276 a former physician and cleric became Pope John XXI, which might have elevated belief in medicine, but Pope John XXI's career progression was highly unusual and his background in medicine had little effect on general practices of the day.

As religious rituals began to become more important, relics—a hallowed bone, a tooth, or a toenail—were used to ward off disease or to bring about cures. (For more on religious healing, see chapter 3, "Diagnosis and Treatment.")

THE IMPORTANT SCIENCE OF GARDENING

After the collapse of Rome, city life throughout the Roman Empire broke down and was replaced by a feudal society, with castles and a ruling class that oppressed the peasant class. Agricultural life dominated, but Roman farming methods were less effective in northern Europe where the soil was less fertile and the climate undependable. As a result, the crops produced in the northern area were not as bountiful. During the early Middle Ages, the peasant diet consisted primarily of starches; very little animal protein was available to them. Hunting was not practiced by the lower class, and domestic animals were not yet raised to provide food. As a result of poor nutrition, peasants suffered from health problems that also led to a greater incidence of disease. Later on, a better plow and a horse collar made farming easier, and peasants increased production of legumes, which offered a protein source that had been absent from their diets. This improvement in eating eventually led to an increase in population.

During this time, people needed to be self sustaining, so the garden was a very important part of the land and the culture; the well-tended garden was a source of food as well as a resource for medicinal plants. If the climate permitted, almost every manor, abbey, and great estate would have had a combination of cultivated lands that included utilitarian gardens, farm fields, woods, orchards, and vineyards. The people of this time were heavily influenced by the farming methods and suggestions of the Greeks and Romans, particularly Columella, who wrote *Res rustica* (On agriculture); Varro, who wrote *De re rustica* (On agriculture); Cato, *De agricultura* (On agriculture); Palladius, *On Hosbondrie* (On husbandry); and Pliny the Elder as well as Dioscorides.

Medieval gardens were always protected in some way. Some gardeners enclosed their spaces with baked brick or stones used to create walls. Others dug ditches or planted trees or hedges to demarcate the boundaries of the garden. Still others wove a wattle (a sort of basket-work of willow branches) that was placed around parts of the garden. While it is doubtful that most of these barriers kept out wildlife, they would have outlined property boundaries and soon become an integral part of a beautiful garden.

The Medieval Garden

During a time when people needed to produce locally most of what they needed, the garden—whether beside a peasant's home or behind abbey walls—was very important. Medieval people excelled in creating the perfect garden, which was an appropriate combination of

Medieval Court Garden

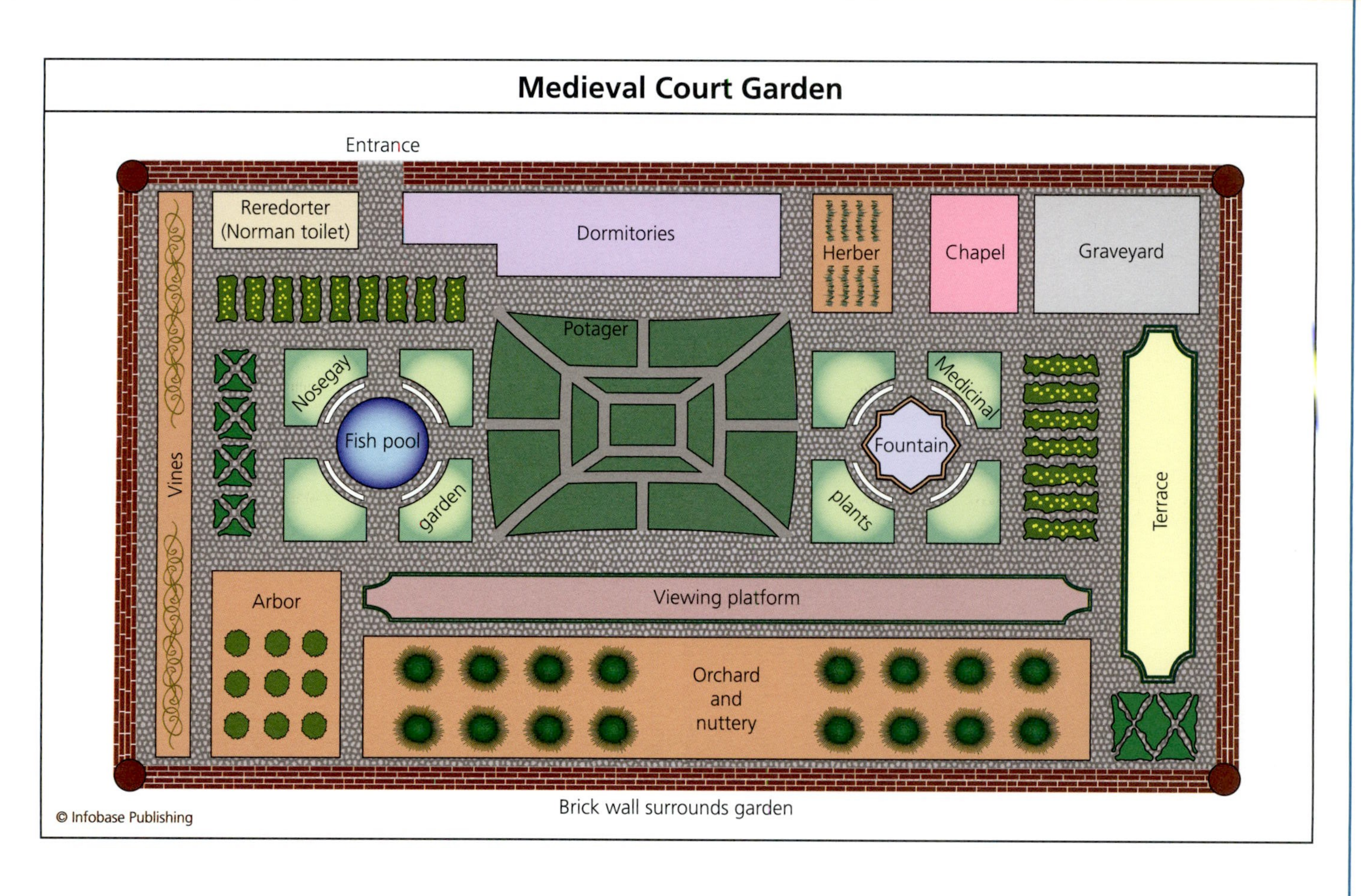

foods to eat and plants that could be used as medicine. They also valued plants for their beauty and their smell, believing that those elements, too, were comforting if not curative. Special gardens were maintained by castles, palaces, monasteries, *hospitals,* and the homes of noblemen and peasants—everyone needed a garden.

The more elaborate gardens contained the following elements:

A potager, which was a formal, geometrically structured vegetable garden that also featured herbs grown in pots. (The term *potage* referred to substances that could be placed in soups to make them thick and nutritious.)

An orchard and nuttery where fruit and nuts could be grown

A culinary border generally surrounded the basic vegetable garden and featured plants that were used to flavor food

A medicinal garden was a vital part of any decent-sized garden. These plants were grown for their medicinal or "physick" use, and included angelica, *mandrake,* henbane, lungwort, foxglove, selfheal, and comfrey. The plants were then turned into healing entities in a variety of ways. They were dried, the seeds were eaten, parts of the plant were mixed into drinks, or they were mixed into a pastelike substance and used as a poultice.

The dyers border was to provide flowers that could be used as ink for writing manuscripts and as dyes for cloth and sometimes mixed in with food or drink.

The more elaborate gardens that were part of royal gardens, estate gardens, and some abbey gardens featured various decorative aspects as well. Some of the elements included the following:

(Opposite) People needed to be self-sustaining, so they learned to grow edible plants as well as plants that could be turned into medicines. Garden space was also allotted for flowers that looked beautiful and smelled good.

A knot garden, which was enclosed by a hedge, and the bushes and greenery were planted in decorative patterns. These gardens were designed to be viewed from above and always featured a viewing platform.

A flowery mead featured more casual plantings that flowered. Nosegay gardens were also popular but were more formal. Both of these types of gardens were generally planted in an area where they could be seen and smelled from nearby buildings; people enjoyed the benefits of keeping wonderful scents and beautiful colors near where people spent most of their time.

Arbors, green expanses of lawn, and statuary were also important in larger gardens. Trellises often lined walkways, and toward the 1500s, *topiary* (plants clipped into fantastic shapes, often that of animals) began to decorate the garden areas. These elements provided beauty and cool places to walk during hot days. Dining al fresco was popular, so tufted seats were sometimes featured. Gardens on estates might have had a menagerie or a simple collection of deer, rabbits, and other wild animals, an aviary with cages containing nightingales, blackbirds, linnets, and other singing birds, a fish pond, and facilities for tournaments.

Water was important for irrigation, so gardens were often built alongside streams, and water was channeled into the garden. The pleasant sound of a fountain and the cool sense of a pool were popular. By the 16th and 17th centuries, royal and estate gardens often featured fountains that could be used to tease, squirting water at passersby at inopportune times.

OTHER SCIENTIFIC EXPLORATIONS OF THE TIME

Astrology and alchemy were two other areas that received a great deal of attention during the Middle Ages. From the time of the Greeks, astrology was used to forecast the life of an infant by observing the alignment of stars and planets at the moment of birth. Medieval healers also paid attention to astrology when it came to treatment of the ill. The zodiac man was one of the most popular illustrations in medical books of the time. The belief that the humoral balance was influenced by astrology became deeply

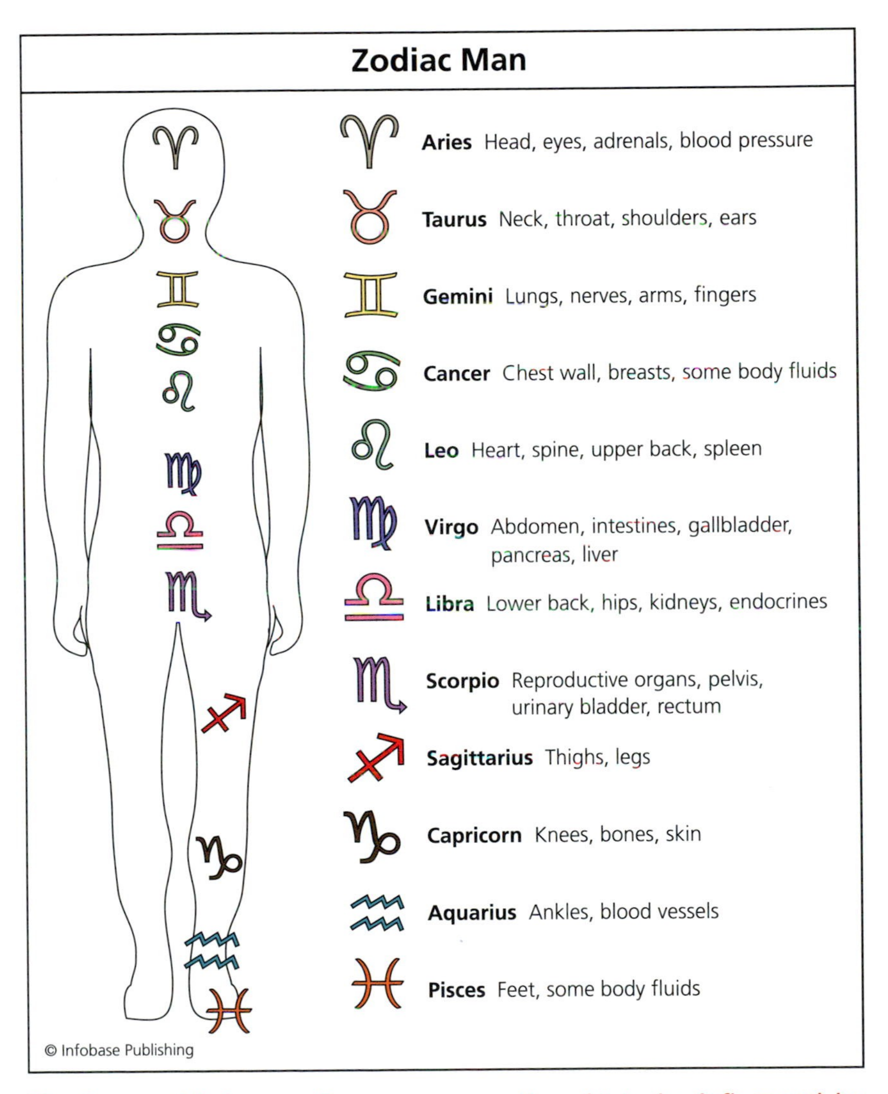

The humoral balance of a person was thought to be influenced by astrology, so the "zodiac man," an illustration that depicted which astrological sign dominated what organ, was one of the reference tools used by physicians and healers.

Canon of Medicine by Ibn Sina (Avicenna) in Latin translation *(University of Texas Health Science Center Library)*

ingrained, and the zodiac man—a graphic depiction of which astrological signs dominated what organs—was an important part of treatment.

Though alchemy is now thought of as the search for a way to turn a base metal into gold, medieval scientists became fascinated by it following the *Crusades* (1095–ca. 1300) when the first texts on this subject were translated from Arabic into Latin. Alchemy was based on Aristotle's theory of earth, air, fire, and water. While the focus was often on gold, those who practiced it could be working on changing any number of substances into something else. Practitioners made certain they were surrounded by symbolic colors, and they recited magical *incantations* while sometimes trying to work out scientific principles.

Over time, any type of science seemed like a threat to the church's authority over life, and eventually Aristotle's books were banned. Because Western Europe was in disarray in the early Middle Ages, it was left to the Eastern region, Islam, to preserve and translate into Arabic the texts of Galen, Hippocrates, and a handful of other Greek and Roman authors. Eventually this information was reintroduced to the West when it was translated from the Arabic to Latin in order that the texts could be made accessible to Western Europe. (See chapter 8.)

CONCLUSION

While it is sometimes written that medical progress came to a standstill during medieval times, this is not totally accurate. The medical practitioners of the Middle Ages did not make great strides in the science of better health and healing, but people still got hurt and became ill, and healers stepped in to provide comfort and attempt cures.

The heavy influence of religion combined with the misguided theories of humoral balance and the fact that the societal norm of the time was "not to question" meant that scientific progress moved slowly. But if progress in any field comes through the proverbial "one step forward and two steps back," then the medieval practitioners may have paved the way for those who followed them by demonstrating what did not work. Certainly by the end of the Middle Ages, the world was prepared for major scientific leaps in many fields.

2

Medieval Healers and Hospitals

During the Middle Ages, basic health care and any curative treatments were primarily managed at home by family members, but when needed, outside help was sought. Early on, there was neither a controlling medical elite nor any particular standard for those who practiced healing. Most healers combined their work with another profession, and few were actually trained in any form of medicine. University training for physicians began to be available to a very few starting in the 12th century. When hospitals began to appear, they were originally rest places for travelers; only later were they dedicated to a place for the ailing.

If a nonfamily member was called in to treat a patient during this time, a sick person in western Europe might have been treated by one of three types of healers: a folk healer, a clerical leader, or a surgeon. A folk healer (also known as an *empiric,* a term used by university-trained physicians to refer to nonuniversity trained practitioners) made up the largest category. These healers learned techniques from each other and through trial and error. Clerical leaders were the second group of nonfamily members consulted for medical advice; for them religion came first, and prayers for healing often trumped any actual hands-on medical care. Surgeons were the final type of medical practitioner. In many communities, surgical duties were assumed by the barber because of his basic knowledge of how to stop bleeding if a person was cut during a shave. In communities where there was enough need, some barbers quit providing haircuts and shaves and devoted themselves to serving as a town's full-time barber-surgeon.

By the 12th century, universities began to offer medical training to a very small group of men. Eventually some level of licensing came into play within the fields of medicine and surgery, and this provided some guarantee of a minimum level of training and education. Even during the latter part of the Middle Ages when physicians began to attend universities, there were actually very few schools, and those that existed tended to graduate only five or six individuals each year. Because there were so few people who emerged with these advanced degrees, few people would have had access to a university-trained physician.

This chapter examines the various types of healers, ranging from folk healers to those who were university trained. Apothecaries, surgeons, and barber-surgeons were also important to healing, and eventually hospitals were helpful, first as places to isolate the sick and only later on to provide better health care.

FOLK HEALERS

In the early part of the medieval period, folk healers were all that existed to help anyone who was sick or injured. The term empiric was used as a pejorative by some during the Middle Ages when referring to folk healers; while the word generally is used to describe methodology based on experience, it was used by medieval physicians to describe those who practiced using skills honed through experience, but without regard to the science or "deep insight" into the problem. Empirics made up the largest group of health providers in medieval Europe, and both men and women were included. Some of these empirics were generalists; others specialized in one type of care or another. (*Midwives,* discussed in chapter 5, would have been included in this category.) Their approaches to treatment varied widely, from first-aid measures (simple cleaning and bandaging) to the use of medicinal herbs or prayers and magic.

Empirics had no formal training. Some likely studied under other healers but many learned medicine through experience, relying on their own instincts. Most of these healers understood and

Catherine of Sienna ca. 1515 by Domenico Beccafumi (1486–1551) *(The Yorck Project)*

practiced the use of herbal folk remedies as well as bloodletting, cauterization, and cupping. The quality of overall care and the results varied greatly. Many would have provided sensitive care that would have been comforting and perhaps effective, and while many had good reputations, written records show that empirics were commonly cited in civil trials and in criminal prosecutions. Many of these cases were filed when a cure failed; very few lawsuits arose over lack of licensing.

Some healers were in the regular employ of a major household or a monastery and earned an annual income. Others were simply the local person to whom people turned for help. Still others were "for hire" on a case by case basis and traveled from community to community. Within this group there were probably charlatans who sold potions and left town quickly.

Later on, these folk healers were excluded from the medical profession when university training became possible for some, and a trend toward licensing requirements began to grow. However, practically speaking, healers still were likely in high demand since most town residents would not have had access to a university-trained physician, nor would they have been able to afford them even if these professionals had been more easily available.

RELIGIOUS LEADERS AS HEALERS

Medieval people were taught that some illnesses were sent by God to punish individuals or communities, and therefore, it only made sense to turn to church leaders for help. While the medical background and methodology of priests varied greatly, these religious leaders offered great comfort to many who were ailing. "Before and above all things, care must be taken of the sick" read the *Regula Benedicti (Rule of St. Benedict)*, written for monks by St. Benedict (ca. 530). (The Rule of St. Benedict is one of the most important written works in the shaping of Western society because it included a written constitution, authority limited by law, and a degree of democracy.) And while basic care was always attended to, prayer was the prime healing method used along with whatever folk methods a priest chose to follow.

Saints were also an important part of religious healing. Certain deified individuals were associated with specific illnesses, and they

Saints Cosmas and Damian—an icon from the 17th century *(Historic Museum in Sanok, Poland)*

were often called upon when other remedies failed to help. The church only recognized miracles by canonized saints, but if the advice of a religious leader led to improvement in a patient, then, understandably, the popularity of that particular healer grew, with the healer taking on saint status to the general public.

In 1130 the Council of Clermont brought an end to the practice of medicine by monks. Council members determined that it was disruptive to religious prayer and the monastic life to have monks tending to those who were sick. As a result, medicine fell into the hands of the secular clergy, and this eventually led to a desire for a more scholastically oriented medicine taught in a university setting.

UNIVERSITY-TRAINED PHYSICIANS AND OTHERS WHO LEARNED FROM BOOKS

The creation of medical training programs in a university setting was one of the most important developments in medicine during the Middle Ages, and the men who trained at universities became the elite of the medical profession. The university movement began in the 12th century with the founding of universities in Paris (1150), Bologna (1158), Oxford (1167), Montpellier (1181), and Padua (1222). In Oxford and Paris, medical instruction was informal at first. By the 15th century (1436), Padua had become the most highly respected of the medical schools, with fifteen medical professors.

Universities were originally founded as places to study Christian theology, and women were barred from becoming clergy in medieval times, so women were not admitted. The exception was in Salerno where women were permitted to teach and practice medicine. Though these women excelled and were well known for their medical skills, their numbers dwindled by the end of the 12th century and the profession came to be dominated by men.

In addition to excluding women, men of the Jewish religion were mostly kept out as well. (A few universities admitted Jews if they were willing to pay higher fees.) Despite this discriminatory practice, Jewish men found ways to become educated, and they often had the finest reputations, which meant they were in high demand by royal and other noble households.

In all universities, a working knowledge of Latin was a prerequisite for admission since the texts and lectures were in Latin. Incoming students also needed basic knowledge of logic and philosophy in preparation for medical courses. This limited those who qualified. Members of the clergy or those who were training to be clergy had the right background because they learned Latin in order to read and understand theology. In addition, a few laymen from relatively wealthy families also had the opportunity to gain an education that prepared them for studying medicine.

The original curriculum in medical schools at that time was based on *The Articella* and Avicenna. (See the sidebar "*The Articella* as a Basis for University Teaching" on page 26.) During the 13th century, the Italian religious leader (later to achieve sainthood) Thomas Aquinas (1225–74) introduced an emphasis on the naturalism of Aristotle in the university curriculum, integrating it with Aquinas's own primary interests in scientific rationalism and theology.

As the educational system evolved and favored the training of clerics for the medical profession, it set up an interesting dynamic: Most physicians were clerics but later on, the clergy in many jurisdictions were forbidden from practicing medicine. While caring for the sick was viewed as a Christian duty, the church became concerned about two issues involving the practice of medicine: The first was that blood was never to be shed by clergy, so clerics had to abstain from any sort of surgical treatment including any form of study that resulted in bleeding. The second issue had to do with financial success. Practicing medicine became lucrative, and the church worried that the clergy would neglect their religious responsibilities for personal gain.

To qualify as a doctor of medicine, a student underwent a full 10 years of training. After this formal education, there was a period of supervised practice that was like an internship, followed by an examination that was conducted by other physicians. If all went well, the physician was licensed to practice on his own. Only a few men successfully completed the course of study each year. Because there were so few graduates, the number of university-trained physicians was exceedingly small, meaning that only nobility or the very wealthy might have had access to these specially trained individuals.

The medicine practiced at this time was a blend of Galen's theories as influenced by Hippocrates. Diagnoses were made by checking the *pulse* and studying the *urine* (uroscopy). Astrological charts were consulted and "critical days" were noted. The prime remedies were dietary changes to rebalance the humors, and the use of herbal drugs. The physicians who served the nobility were known as court physicians, and their role was actually a broad one. Records show that a court physician to Henry III of England was expected to counsel the steward about the choice of meat and drink served to the king. He was also to be on watch for signs of pestilence or the arrival at court of anyone who might have leprosy, and the king was to be warned of signs of either of these threats.

The ratio of physicians to townspeople in Florence, Italy, in 1338 provides an interesting picture of the time. Florence had the highest physician per capita rate with about 60 licensed physicians for a population of about 120,000. After the Black Death, the ratio improved because of the precipitous drop in population: There were then 56 physicians for only 42,000 people. A few towns paid for physicians to oversee the care of the public and treat the poor, but for the most part, physicians treated those who could afford to pay. By 1211, the town of Reggio, Italy, not only paid a public physician, called a *medici condotti,* to help care for the poor, but they expanded the physician's duties to include helping with inquests, treating the sufferers of plague, and tending to injuries inflicted on prisoners.

"THE ARTICELLA" AS A BASIS FOR UNIVERSITY TEACHING

The Eastern Roman Empire preserved a good number of medical texts from Greek and Roman times, where they had been rewritten in Arabic in order to make them more widely available throughout the Islamic Empire. One monk became interested in translating these medical texts into Latin, and his efforts helped launch an interest in scholarly training for physicians in the West.

Constantinus Africanus (1010–87) was a native of Carthage, under Arab rule at that time, so Constantinus was exposed to a variety of languages. His

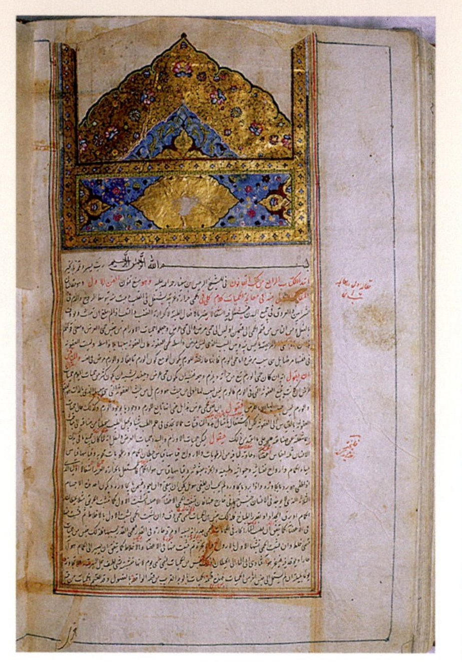

Islamic Medical Text c. 1500. Illuminated opening of the fourth book of the Kitab al-Qanun fi al-tibb (The Canon on Medicine) by Ibn Sina (Avicenna). Undated; probably Iran, beginning of 15th century *(National Library of Medicine)*

OTHER HEALERS: APOTHECARIES

Healers tended to rely on herbs and common ingredients that they could grow or easily gather. More advanced medical professionals often requested ingredients that had to be imported, and for which they turned to apothecaries, the forerunner of today's pharmacist. *Apothecaries* always maintained full stocks of the more popular herbal cures and spices, such as saffron, background made him attractive enough that he was invited to base himself in Italy at the Schola Medica Salernitana (Salerno). Many of the works of Greek and Roman philosophers and physicians, including Galen and Hippocrates, had been preserved by the Islamic countries and translated into Arabic. Constantinus was able to translate much of this work into Latin to make it accessible for the first time to western Europeans, and so westerners began to study the medical methodology recommended by these men. As a result of this influence, Galen, who had written on many subjects, continued to dominate medicine for the next few centuries. Other monks eventually contributed to what Constantinus was working on, and the professors at Salerno began using this canon of writings, known as the *Ars medicinae* (art of medicine) or *Articella* ("little art of medicine"), as the basis for European medical education for several years.

Later Gerard of Toledo, Spain, (ca. 1140) translated hundreds of works by Aristotle, Ibn Sina (Avicenna), Razi as well as Abu al-Qasim's (Albucasis) writings on surgery. The *Qanun* of Ibn Sina became the cornerstone textbook of medicine at the University of Montpellier, the largest nonclerical institution in Europe until 1650.

cinnamon, and ginger, and they often imported specialty items that were thought to be curative. These included ivory, gold, pearls, mummy dust, and ambergris (from whales). Sugar was a key ingredient for improving the flavor of medicinal syrups and pills. Some apothecaries created a sideline by selling sugar pills to wealthy clients who liked sweets, creating an early form of candy.

In addition to importing ingredients, the apothecary was also responsible for mixing them into ointments, pastes, syrups, and pills. Sometimes they were custom made at the instruction of a doctor; other times the doctor came and supervised. As they are today, some medicines were "over the counter" and some were "by prescription." In some jurisdictions, the apothecaries' wares were subject to periodic inspections by the local physicians. The physicians were given the authority to have the apothecaries toss out anything found to be stale, watered down, or in any way defective. Tensions sometimes arose when physicians accused apothecaries of diagnosing and dispensing medicine that should have only been prescribed by a doctor.

EARLY SCIENTISTS

Albertus Magnus (ca. 1200–80) was a priest born in Germany, who was educated at Padua and went on to become a fine teacher in Paris. While a certain degree of science had been used during the Roman Empire, Albertus was the first European to use the scientific method. He built his theories on Aristotle and added to these the idea that there could be the possibility of chemical change. The church ultimately canonized him as a patron saint of scientists.

Thomas Aquinas (1225–74) was regarded as a great theologian and philosopher. He became the principal interpreter of Aristotle's works, and he taught a methodology that combined Aristotle's emphasis on naturalism with scientific rationalism and theology. Aquinas was canonized in 1323 and came to be considered the patron of Catholic schools.

Roger Bacon (ca. 1220–92) was a Franciscan monk who explored the world of science under the protection of Pope Clement IV. When Clement died, Bacon was put in prison and some of his writings were burned. In 1733, 450 years after his death, his major work was finally published; it revealed that even earlier than Leonardo da Vinci (1452–1519), Bacon predicted steamships, automobiles, submarines, and flying machines. In 1230 he wrote of the possibility of the world being circumnavigated. One of his medical creations was the invention of spectacles.

False Gerber was a Spanish monk who was interested in chemistry. (The original Gerber was an Arabic mathematician and chemist whose real name was Jabir ibn Hayyan (ca. 721–ca. 815) so when another came along, he became known as False Gerber.) False Gerber is now recognized for making the study of chemistry into a modern science. He discovered vitriol (sulphuric acid), which was the greatest chemical advance since the discovery of iron smelting 3,000 years earlier. False Gerber also described how to make strong nitric acid, a process that was originally created by alchemists. His writings date from about 1300.

Arnold of Villanova was a physician born in Spain in the 14th century whose work focused on improving the elixirs that were being used medicinally. Alcohol was among them; Arnold learned to isolate it and called it aqua vitae ("water of life"). He also came to understand the existence of carbon monoxide.

MEDIEVAL HOSPITALS

Hospitals in medieval times were not really hospitals in the modern sense of the word. While some offered minor medical care, they were primarily created to offer food and shelter to travelers ("hospital" is derived from the Latin word *hospes,* which means guest or foreigner). Religious entities were the primary sponsors of these facilities. A Christian organization, the Knights Hospitaller (later known as the Knights of Malta), established many hospitals during the 12th century in areas through which people traveled on the Crusades so that soldiers and pilgrims would have a place to rest and get food. Over time, the Knights of Malta began to expand its offerings, establishing facilities elsewhere in Europe.

Almost half of the hospitals in medieval Europe were connected with monasteries, priories, or churches. These facilities generally started out as infirmaries for the care of monks and nuns, but eventually they broadened to offer aid to anyone who needed it, with monks and lay helpers providing a basic level of health care,

Crusades, the Massacre of Antioch, by Gustave Doré (1832–1883)

regular meals, and rest, as well as spiritual guidance. Few hospitals had physicians though a motivated cleric might have studied some treatments described in texts in the hospital library.

A February 1992 article from the *Journal of Public Health* quotes from an account of life at the Hôtel-Dieu in Paris: "Nurses arose at 5:00 A.M., attended chapel prayers after ablutions, and then began work on the wards. Their duties included using a single portable basin to wash the hands and faces of all patients, dispensing liquids, comforting the sick, making beds, and serving meals twice daily. Sisters on night duty reported at 7:00 P.M. It was their task, in an era before the bedpan, to conduct the ill to a communal privy, for which purpose the hospital provided a cloak and slippers for every two patients."

The facilities that were paid for by charitable donations (alms) were sometimes referred to as almshouses (a term for poorhouse today). In return for food and shelter, people who stayed there were expected to pray for the souls of the donors.

The medieval hospitals of the day generally turned away people with infectious diseases. The standard care consisted of relatively clean surroundings and nourishing food, with an occasional dose of something from the apothecary. Hospitals were not used as teaching facilities in the Middle Ages, although there are a few mentions of surgeons offering instruction in a hospital setting.

Some hospitals that began as small additions to monasteries became major institutions by the 12th century. Two hospitals in London that became well known were St. Bartholomew's and St. Thomas's. In Italy, Milan had about a dozen hospitals, and by 1400 Florence had more than 30 hospitals, the largest of which was Santa Maria Nuova. By 1500, this hospital had a staff of 10 doctors, a pharmacist, and several female surgeons attached to it. Over time, the hospital movement spread throughout Europe. Every city in Germany with a population of more than 5,000 had a hospital, and the Norman invasion of England also brought the movement there. Sizes of these facilities varied. Some accommodated only 10 to 50

Medieval hospital. "La Grand Chambre des Povres"—the Great Room of the Poor—is believed to be the world's oldest edifice to have been in continuous use as a hospital. Representative of medieval hospitals, it is a part of the Hôtel-Dieu de Beaune, France, founded in 1443. Combined with modern professional hospital service, it carefully preserves the atmosphere of the 15th century. Sisters of the Congregation of Sainte Marthe, garbed in habits traditional to their ancient order, have cared for the sick, the aged, and the indigent in this hospital for more than 500 years, uninterrupted by wars, or by economic or political changes. *(Department of Library Sciences, Christian Medical College—Vellore, History of Medicine Picture Collection)*

Hôtel-Dieu, between 1890 and 1900 *(Library of Congress, Prints & Photographs Division)*

people but others were much bigger. St. Leonard's in York could care for as many as 225 sick or poor individuals by the late 13th century, though this was still small compared to today's standards.

The interior design of most hospitals was like a big residential dormitory with a chapel, a kitchen, and laundry facilities attached. If the facility serviced both men and women then there were generally two separate wings to house them. Typically, more than one person was assigned to each bed. This was common practice in households, so patients probably didn't think much about it. In one account, 12 children were expected to share one bed. By the 15th century some facilities were moving toward private beds with curtains for privacy. The wealthy would have been kept at home if they were sick, or grander arrangements would have been made for them if they were traveling, so these facilities were primarily for the poor.

With continued leadership from the church during the 13th century, the hospital facilities became bigger and better designed.

Hospitals were generally situated in an area with a peaceful, airy feeling, and some of the buildings in Italy were quite beautiful. One hospital had a wing added that was designed by architect Donato Bramante (1444–1514) and another by the artist Michelangelo (1475–1564).

A good number of hospitals offered specialized care. Some were created specifically to provide longer-term help for the poor, or for the blind, the lame, the elderly, or the mentally ill. There were also facilities for unwed pregnant women as well as for babies who were abandoned or whose mother died at birth. (Though hospitals tried to send foundlings out to wet nurses, if such could not be found the staff had to feed the infants with cloths dipped in milk.)

Lazar houses (after the diseased beggar in one of Jesus' parables, also known as lazarettos) were refuges for lepers who were forbidden from living within any regular town or community. Near Canterbury, England, a large facility for lepers was built, and more than 100 lepers lived in scattered wooden buildings. The

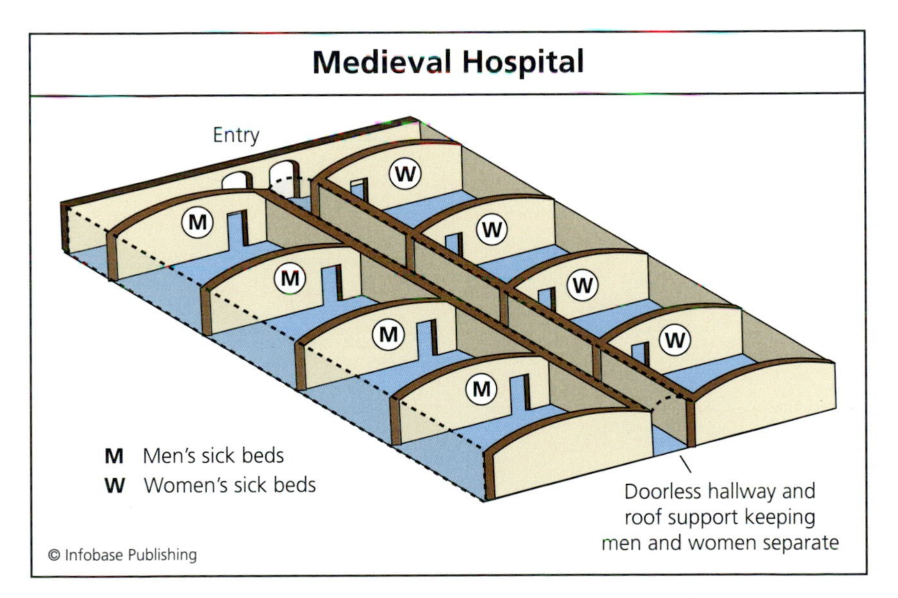

Medieval hospitals were set up like large dormitories. Men and women were separated but frequently more than one person was assigned to the same bed.

isolation was felt to be important to keep the disease from spreading to healthy people in the community; however, the lepers were given fresh food and had access to water that was thought to be medicinal. Once admitted, they did not expect to leave. Unfortunately, some cases that were diagnosed to be leprosy were actually just unsightly skin conditions that eventually healed. But once admitted to one of these hospitals, one's fate was sealed.

CONCLUSION

Medieval healers ranged from folk healers who may have been quite good at their work to university-trained physicians whose education was based on Galen and Hippocrates, with few improvements since the origins of these men's theories. Understanding of anatomy was inaccurate, and while medieval practitioners were somewhat successful at divining helpful herbal cures, the medications and the dosages were all developed through trial and error. It would have been a dangerous time to be a patient.

Hospitals began as places offering food and rest for travelers. Most were attached to or run by a religious organization, and over time, these facilities began to offer some basic medical care.

3

Diagnosis and Treatment Methods

Medieval medical healers and physicians took diagnosis and treatment of their patients very seriously. An initial diagnosis involved evaluating a person's normal state of health, taking into account his or her symptoms, assessing (nonscientifically) the person's blood and urine, and then considering how any treatment would be affected by the current position of the stars. Other than checking a patient's pulse, a physician would rarely touch a patient.

Basic first aid was important in helping a patient; dietary changes were sometimes recommended, and bloodletting and cautery (see chapter 4, "Surgery in the Middle Ages") were also very popular "cures." However, herbal remedies or religious healing were generally the preferred solutions. Herbal remedies were mixed according to the specification of the healer or the physician (sometimes mixed by the healer him- or herself). Sometimes the herbal remedies were to clean the body (through purging or evacuation of the bowels) in order to rid the person of a particular illness. Surgery was high risk, so while a growth might be removed or a bone set, invasive surgery was avoided whenever possible.

This chapter will examine the methods physicians and healers used to diagnose an illness and then will explain the way medications and religious healing were employed. Mouth pain and mental illnesses were also serious problems of the time, and how these issues were managed will also be discussed.

DIAGNOSTIC METHODS

The medieval medical practitioner felt that the key to diagnosing a person's ailments was assessing the person's humoral balance to determine the nature of the illness and to prescribe the appropriate cure. This diagnosis rarely involved physical examination of the patient, which was considered unnecessary, and of course, male practitioners were prohibited from examining a female patient.

Observing the patient was the first step in making a diagnosis. This permitted the physician to identify the person's normal temperament, so that the physician could then identify whether the person had departed from normal behavior and what would bring the person back in balance. (See the sidebar "Identifying and Treating a Patient's Illness" on page 39.) Part of the humoral imbalance needed to be considered in the context of the person's zodiac sign. Physicians generally carried with them a small handbook that contained helpful charts, including an astrological chart to make a proper diagnosis. The two items that needed to be considered next were the patient's urine and blood.

Though medieval practitioners had no understanding of the chemistry of these substances or exactly what they were looking for when they collected them, they believed it was important that these two substances be evaluated. A uroscopy, known then as a way to evaluate the health of the *liver* by checking the urine, was such an integral part of medieval medicine that the clear glass beakers used to hold and then examine the patient's urine became a symbol of the medical profession. Medieval textbooks ran full color illustrations of the 20 recognized colors of urine. The urine colors were displayed in graph form and ranged from colors signifying illness (blue, black, and dark red) to those that were viewed as healthier, including several shades of yellow right through to almost clear. Physicians considered smell and texture (thin and watery or thick and greasy), and they looked for any sediment or other solids in the urine. A medical writer from the 13th century noted that thick urine that was whitish or bluish-white indicated "dropsy, colic, the stone, headache, excess phlegm, rheum in the members, or a flux."

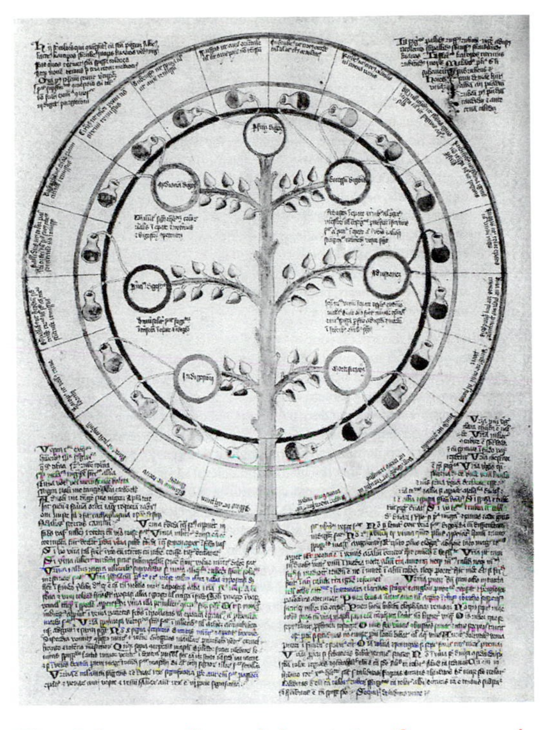

Physicians evaluated the state of a person's health by comparing a person's urine specimen with a urine color chart.

Some of the elements noted are ones that are still evaluated today: Red in the urine was understood to indicate blood in the urine, and grit that was visible in a urine sample might lead the practitioner to consider the presence of kidney stones. As with blood analysis, the results were interpreted along the theory of humoral balance.

Physicians sometimes performed bloodletting and observed the strength of the flow and the speed of clotting afterward to analyze imbalances. The blood was often collected in a bowl or a cup, and they studied it for its odor, its warmth (or lack thereof), and its texture.

Checking the pulse was also important. Physicians realized it reflected on the state of the heart, and there were many classifications of what the various conditions of the heartbeat meant. They learned to evaluate the pulse for duration, breadth, strength, and regularity of beat. In the 13th century, physicians wrote of distinguishing among the following five considerations:

- motion of the arteries
- condition of the *artery*
- diastolic and systolic duration and pressure

This is a 15th century painting known as *Les très riches heures du Duc de Berry.* Signs of the zodiac are depicted in their correspondence to each part of the body, starting with Pisces, the feet, and working up to the head, Aries. *(Musée Condé, Chantilly, 15th century)*

- strengthening or weakening of pulsation
- regularity or irregularity of the beat

Physicians were carefully trained to evaluate the nuances of these elements.

After considering the necessary factors, the physician would place these symptoms within the context of the patient's astrological sign. The healer then felt prepared to identify the disease and predict its course and outcome. Of course, this was merely a best guess but it was all they had at the time. Physicians prided themselves on their accuracy, particularly in regard to whether or not the patient pulled through. A physician's reputation rose and fell based on the accuracy of the predicted outcome.

TREATMENT: POPULAR MEDICINES

Medicines were used by both empirics and university-trained doctors as part of most treatments. Medications were compounded by the medical healer or physician, or later on, by the local pharmacist from ingredients kept in an apothecary. Trial and error was very much a part of the treatment process. The results were mixed, ranging from effective (there was some knowledge of good pain killers and laxatives) to harmless (similar to taking a sugar pill with possible improvement from the psychological expectation of getting better) to dangerous (some remedies—or their dosages—were actually poisonous) and disgusting (pig manure was thought to be helpful for nosebleeds).

Herbs and Vegetable-Based Ingredients

The most common medicines were made from plant matter and herbs. Herbs ranged from the familiar, such as rosemary, sage, marjoram, mint, and dill to lesser known ones like squill, pimpinella, henbane, betony, and pennyroyal. They were mixed in many different ways and were prescribed for common ailments such as nosebleed, baldness, sunburn, and loss of appetite. Some were made into drinks mixed with ale, milk, honey, and vinegar.

IDENTIFYING AND TREATING A PATIENT'S ILLNESS

The medieval physician's job was to identify what was out of balance in a patient in order to diagnose and treat the illness. The cure always involved righting the balance. For example, stomach discomfort was thought to be caused by an excess of phlegm in the lower digestive tract, so the remedy needed to be a substance that would help rid the patient of excess phlegm. (This was generally referred to as colic, the term used to describe all types of stomach ills of the day.) An imbalance of the phlegmatic humor was associated with things cold and damp, so the remedy, according to medieval practice, was warm, dry food. Dill and the meat of a rooster were particularly recommended. As it happens, there was some wisdom in this solution: Dill actually provides relief for gas pain, and simply prepared poultry is a soothing, nourishing food. A patient with colic was also told to avoid foods like fish and oranges, both of which were categorized as cold and wet and would not have been as easy to digest.

The roots of the mandrake plant were highly valued as a painkiller but the plant grew only in the Mediterranean region. The added cost of shipping it out of the southern region made it a very expensive medicine that only wealthy people could afford. Hemp was eventually identified as a cheaper alternative to mandrake. Every manor grew hemp for its fibrous stalks that made good ropes, so it was readily available for use as a painkiller at a much lower cost. The poppy plant (from which *opium* is made) was also grown as a painkiller; sometimes mothers used poppy juice on their nipples so the baby would sleep after feeding. Cinchona bark (quinine) was another remedy of the period, and this plant is still viewed as effective for malaria.

Little has been written about how they determined food qualities, but as with herbal remedies, the medieval food values were almost certainly a combination of association (what the food reminded them of) and trial and error. For example, fish are cold and wet, so they were identified as bringing out the moist and cool humoral balance. Experience taught medieval people which foods were calming to the stomach and which were not. Over time this information was codified, so charts and graphs were created to provide healers with lists of foods and their various qualities. The methodology behind these beliefs was rudimentary and not very accurate, but it shows a basic understanding of the effect nutrition could have on a person's health.

Astrological predictions were very important to any type of treatment of the time, so even when recommending a lifestyle change, a skilled physician also factored in the position of the stars and what effect it might have on what the patient ate.

Spices were also thought to be curative. Cumin, cardamom, ginger, cloves, and many others were part of recipes for medications. Vegetables such as rhubarb and lettuce were commonly added, as were various vegetable oils made from certain plants or their seeds.

Monasteries and royal families grew extensive gardens filled with herbs and spices that could be used in cooking as well as for medicines. (See the sidebar "Types of Gardens" on page 43.)

Nonvegetable Ingredients

Some elements that were used as medicines were more like what might be found in a "witches' brew." One medieval physician recommended applying pig dung to a patient's nose to stop a nosebleed. Raven droppings were listed as an ingredient in an ointment to relieve toothaches. Teeth, fat, and the grease of animals were also used. Some ingredients were very costly to obtain; among these were gold, powdered gemstones, and ambergris (an opaque, ash-colored secretion of the sperm whale intestine, usually found floating on the ocean or cast ashore). The more exotic the ingredient, the more likely it was that the medicine was being created for someone wealthy. Mercury was another highly prized "cure," and no one realized how very dangerous it was until the 20th century.

Another popular medication, originally touted by Galen and important throughout the Middle Ages, was theriac, made from vipers' flesh and a multitude of other ingredients. It was extremely popular for the wealthy and was used for everything from epilepsy to chest constriction and smallpox. Depending on the preparer, theriac was assembled with at least 64 ingredients, including goat dung, pieces of mummy, and adders' heads as well as vipers' flesh, opium, and a long list of special spices and herbs; cinnamon, mushrooms, fungi, and gum arabic were among them. The brew had to be boiled and left to ferment for several years, and it was eventually reduced to a pasty substance that could be mixed into a salad or a drink. As with other medicines that were boiled, the very act of heating the mixture for a time meant that some of the impurities were destroyed, thereby reducing some of the dangers of the ingredients.

HOW MEDICINES WERE CONSUMED OR APPLIED

Medications were usually consumed as some form of liquid (a syrup or a cordial), sucked on as a lozenge, or sometimes eaten or swallowed in a pill-like form. If the medicine was easy to make as a powder, then it could be dissolved in another liquid or mixed into a paste to be used topically, called an electuary.

If the medicine needed to be taken by mouth, then it was generally sweetened using honey or sugar, and sometimes the pill was not much more than a simple sugar pill. At least sugar pills would

TYPES OF GARDENS

Gardens were vital to rich and poor during the Middle Ages. Importing anything from afar was difficult and costly, so food as well as herbs and spices used in medicines primarily had to be home grown. The following types of gardens created were based on their primary purpose:

Peasant gardens. These gardens mainly featured vegetables with as many medicinal herbs as peasants could grow successfully. They were generally surrounded by wattle (woven) fences in an effort to keep pigs out of the garden.

Abbey gardens. Monasteries had multiple gardens, including vegetable gardens and "herbers" for cultivating foods and herbs to eat, an infirmarer's garden of medicinal herbs, and orchards created for growing fruit but also laid out in such a way that monks came to them to both pray and pace (monks frequently pondered difficult questions while pacing). Monasteries realized the benefits of wonderful smells, and the plans for most abbey gardens show that fruit trees were planted near the *infirmary* so that people could smell the wonderful odors. Abbeys also sometimes set aside land for townspeople to grow what they needed.

Gardens for royalty and noblemen. Like monasteries, the homes and castles of the wealthy were multipurpose and extensive. While growing food and medicines was vital to the well-being of those who lived on the estate, beauty and pleasure were also considered in the design. From

(continues)

(continued)

bordered walkways and fountains to "nosegay" gardens that gave off a beautiful fragrance, the ideal medieval garden was both practical and beautiful. (Also see chapter 1, "Medical Beliefs in Medieval Times.")

Infirmary/hospital gardens. These gardens were used to grow herbs, fruits, and other plants for the relief and refreshment of those who came to the hospital. Much of what was grown was used in cooking and for relief through the preparation of medicinal baths and for other health-giving purposes. In smaller hospitals, the sisters took care of the gardening in addition to tending to the patients and visitors. In some communities, the gardens provided enough produce that surplus apples, pears, onions, and leeks could be sold on the open market as a cash crop. Most hospitals also created "paradyse gardens" to offer solace to the troubled or grieving. While cultivation of these hospital gardens was mainly performed by women, in wealthy areas, a gardener did the work but took orders from the matron as well as the physician and the surgeon. In English almshouses, the more mobile and capable residents were expected to help with tending and weeding the gardens.

Gardens helped make the medieval people self-sufficient in growing their own food and developing knowledge of medicinal plants. Later on, the science of botany became very important to the development of medications.

have been harmless, and perhaps patients experienced a placebo effect when they were told that a medicine would improve their symptoms.

In certain cases, medications were thought to work better if they were inhaled. In this case, the liquid medication was applied to a sponge or cloth that was then held over a patient's mouth and nose so that the vapors from the medication were taken in through deep breaths. This was common with anesthesia-type medications. There was also a belief that an *anesthetic*like effect could be delivered to a patient via a process of "fumigation," which involved burning a substance so that the patient would breathe in fumes that would take away the pain. If the surgeon was unaffected by the fumigation, then the patient probably got no pain relief either.

RELIGIOUS HEALING

Medical knowledge of the day was inadequate for solving many health problems, so healing through prayer and divination seemed like a logical option. Since physicians and healers could be expensive and were not available in some communities, faith healing offered advantages. Anyone could pray for help at any time, and every community had religious leaders to whom they could turn for this type of healing.

While the Middle Ages are sometimes referred to as the "age of faith," there is no evidence that prayer was used exclusively in the search of a cure unless a family was so poor that they couldn't afford other care, or unless a patient was so ill that no medical professional—who always had to keep in mind his reputation for care—would take the case.

The wording of prayers from that time shows that the requests made were indirect. A prayer was sent to a specific saint who was then asked to contact God on behalf of the person who was ill. The Virgin Mary was a popular recipient of these prayers, but the appeals were often made to others as well. St. Apollonia was the patron saint for toothaches. Eye problems were addressed through St. Lucy. Patron saints were often martyrs, and the connection

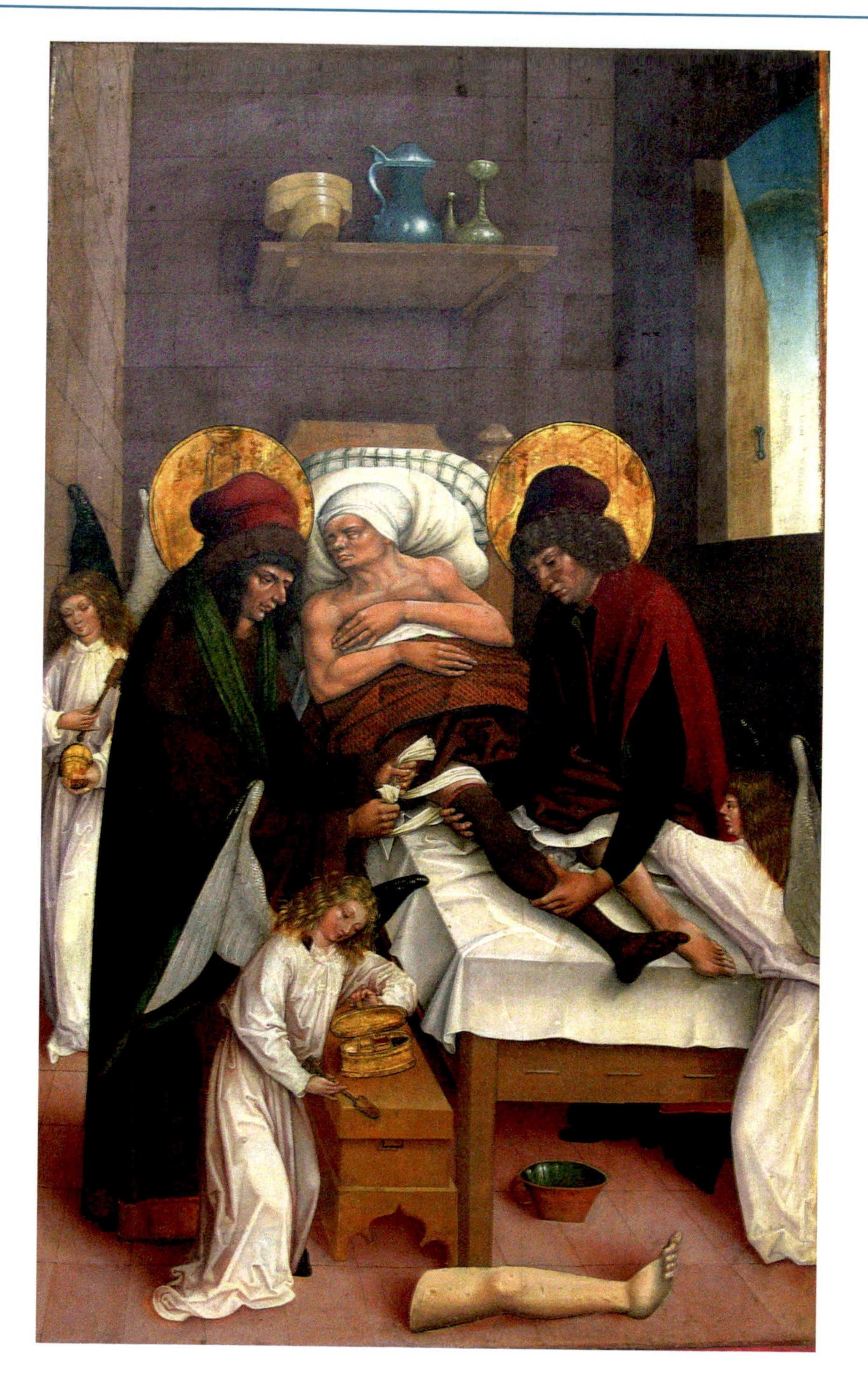

between a saint and a particular condition was often based on some aspect of the saint's suffering. The story was that St. Apollonia had her teeth knocked out and her jaw broken before she was executed, so sufferers of toothaches felt she would be especially sympathetic to what they were enduring.

Religious rituals often took on aspects of the occult. Because people lacked education, the words and phrases used often slurred into more of a magical hodgepodge of nonsense.

Relics

In faith healing, relics were often important to the process, and if a possession that was thought to have belonged to a saint seemed to cure effectively, then myths about the objects grew. This success brought great attention to the house of worship that claimed ownership of the relic, and that church soon became the site of mass pilgrimages, which brought an increase of attention and income. Pilgrimages brought generous donations, and so churches could create shrines decorated with gems and gold, which heightened the reputation of the saint honored. A good example of this was the Thomas à Becket shrine that was located in Canterbury Cathedral. (Geoffrey Chaucer's classic *The Canterbury Tales* was about a pilgrimage there.) The shrine was destroyed during the time of Henry VIII (1491–1547).

Some made pilgrimages to shrines as a sign of devotion to God; others did so to give thanks for a positive development in their lives; still others came seeking a cure. A more arduous trip was thought to give the pilgrim added leverage when making his or her plea. Many added hardship to their journey by walking instead of riding horseback, or making as big a financial gift to the shrine as they could afford, with the hope of being rewarded.

(Opposite) Legendary transplantation of a leg by saints Cosmas and Damian, assisted by angels. Cosmas and Damian are regarded as the patrons of physicians and surgeons and are sometimes represented with medical emblems. They are invoked in the "Canon of the Mass" and in the "Litany of the Saints." *(Andreas Praefcke)*

In addition to cash, some people brought wax images of the afflicted person or of the diseased body part. These effigies were temporarily left out as decorations but because wax was so valuable, church leaders eventually recycled them by melting them down as candles. Churches and shrines also held cast-off crutches and other aids left by those who were miraculously able to walk again or were in some other way cured following a visit to a shrine.

Magic

Empirics relied heavily on magic charms and incantations or spells to try to drive away illnesses. University-trained physicians occasionally used magic too, particularly that which was based in astrology.

During medieval times, people believed that elves and goblins filled the air with invisible powers of evil, so healers would often create *amulets* that were filled with herbs, stones, and other materials as they were believed to generate some supernatural power for healing or for warding off illness. It was believed that this use of amulets could preserve one's eyesight, cure lunacy, prevent one from fatigue while traveling, or even protect one's cattle. The herbs could be worn or hung from a door (usually with red wool), and during pre-Christian times in Rome, a practice developed of wearing brooches and pins that were inscribed with protective words. Because this practice began before Christianity, the words often evoked divine assistance from a god or goddess. This practice survived even after Christianity was prevalent but saints' names were then substituted for the gods that had been used previously.

Incantations usually needed to be used in combination with herbal medicines or a charm. And sometimes it was very important that certain incantations were recited while collecting ingredients for a medication. Additional incantations were later said as the medicine was administered. Many of the incantations that survive in written form appear to be gibberish. Because few people were literate, they were passed down orally, and as a result, it is hard to know how these "magic spells" started out. An incantation may have originally been in Latin but through repeated mispronunciation it may have been reduced to meaningless syllables. Or some may have been based on a Christian prayer but then were distorted over time.

MOUTH PAIN AND TOOTHACHES: DIAGNOSIS AND TREATMENT

Toothaches and other mouth complaints were common among the medieval population, and carvings in churches show people with their jaws bandaged indicating that it must have been a significant problem to many. Poor dental hygiene and a lack of understanding of the importance of certain nutrients in the diet contributed to medieval people's various levels of tooth pain and health ills related to their mouths. Well into the 18th century, dental problems were thought to be caused by worms burrowing through the teeth and gums, so most efforts to improve mouth health were misguided.

Diet issues of the day increased problems with the mouth and teeth. Medieval people did not eat much in the way of sweet or sticky things because sugar and honey were hard to obtain, but their regular consumption of coarse breads caused excessive wear on the teeth. During the winter and early spring, medieval people would have had difficulty finding fresh fruits or vegetables to eat. This led to a deficiency in vitamin C, which leads to scurvy, a disease that is characterized by swollen, bleeding gums that make tooth loss more likely.

The first line of treatment for tooth or mouth pain at this time often involved balancing the humors, generally by carrying out remedies that directly related to the mouth. Sometimes the patient was to chew or eat certain herbs; at other times, cautery or bloodletting was used.

When practitioners spotted holes in the teeth (cavities), they experimented with filling them, often using lead, which created another type of health issue (the very real likelihood of lead poisoning). Teeth were sometimes capped with gold, but filling a tooth was not very effective until acceptable amalgams were invented in the late 18th century.

If tooth pain became too great, the sufferer sought out a barber-surgeon or a traveling tooth-pulling specialist. These healers sometimes poured acid into the afflicted tooth in an effort to halt the pain. This destroyed the nerve of the tooth, which reduced the discomfort, but it usually resulted in damaging the tooth to the point that it could no longer be used for chewing, after which it generally needed to be pulled. Tooth extractions were also problematic. The person's jaw sometimes was broken in an effort to pull out a deep-rooted tooth; other times the tooth came out but the area soon became infected which led to more pain. Even when pulling the tooth brought an end to long-term suffering, the tooth-pulling process itself must have been excruciating with no suitable way to numb the mouth.

Once a tooth was removed, the person was simply left with a gap. Both the Egyptians and Romans had developed bridges and false teeth that could be inserted, but this did not seem to be popular during the Middle Ages.

Dental Hygiene

Even at this time, perfect white teeth and sweet breath were the ideal, but medieval people did not know much about dental hygiene or how to guard against tooth decay. Toothbrushes did not exist until the 19th century, and the concept of flossing to rid the mouth of bits of food did not come about until the 20th century.

Health manuals talk of picking food out of teeth (but it was not to be done in public). Chewing on mallows, a reedy plant that grows in marshy areas, and rinsing one's mouth with wine or vinegar was also recommended. These methods would not cure gum or dental disease, and deficiencies in diet caused additional difficulties.

In the 10th century an Arab physician revived a practice of tooth scraping (which resulted in the removal of tartar) that was tried as early as the 7th century in Greece. This was used occasionally throughout Europe by the 12th century, but it would be a long time before it became a common practice. Despite the difficulties in maintaining dental hygiene, most people tended to keep enough of their teeth, enabling them to eat solid foods for most of their lives.

MENTAL ILLNESSES: DIAGNOSIS AND TREATMENT

There was little understanding of mental illnesses, either the contributing factors or the manner in which they might be treated. Mental conditions were generally thought to have physical or supernatural causes but as with physical ailments, the first method of approach by the healer was to balance the humors. For example, depression was thought to be caused by an excess of black bile (categorized as cold and dry) so physicians recommended diet and a regimen that would provide warmth and moisture. If this approach was ineffective then the physician passed the case on to someone who could evaluate whether the astrological alignments of the person were being affected by demons. If these cures didn't help, then the family's only hope was to turn to someone who could seek help from God, and this made penance, prayer, and sometimes exorcism the cures of choice.

In the 14th century, treatments were often based on superstition, and some of the recommendations of this time involved bathing in human urine, wearing excrement, placing dead animals in the home of the sick person, using *leeches,* and drinking molten gold (gold heated until it melted) or powdered emeralds (a green jewel).

People today who suffer learning disabilities or mental retardation would have been classified during medieval times as "simpletons," "lunatics," "idiots," or "insane." Because it was difficult to pay for and support a person who could not work, the options available to families were few. A relatively well-to-do family might try to take care of the person at home, or sometimes they paid for the person to be kept at a hospital. A few hospitals took in the poor but often the families had to resort to leaving the mentally disabled person on the streets to beg; there were no other options.

CONCLUSION

Diagnosis and treatment during the Middle Ages was well intended but unscientific. The physician usually did not actually physically touch a patient, except perhaps to check pulse rate, and symptoms, normal behavior, and the positioning of the stars were factored into the recommendation of a remedy. Cures ranged from suggested dietary changes to specific herbal medications, bloodletting, and sometimes the use of prayer. The results of healing methods used were entirely mixed. While some of the herbal medications were somewhat effective, luck was still a major factor in whether or not a person got well.

4

Surgery in the Middle Ages

The Middle Ages varied only a little from other times when it came to illness and injury. People were hurt in battles, they were bitten by animals, they fell and broke bones, or they dislocated joints (usually the shoulder). Some suffered from *hernias* or kidney stones. These situations frequently caused great pain and surgical intervention was sometimes the only solution. As a result, those with surgical expertise found themselves in great demand.

Despite this great need, the practice of surgery was viewed poorly by physicians of the time. Surgery was considered manual work, more comparable to the tasks of a tradesman than to that of a physician. Unlike surgeons today who must train and be licensed as medical doctors, surgeons of medieval times were not educated as physicians. It was believed that true doctors healed by observation and by prescribing remedies rather than through any type of physical intervention. This meant that surgeons came to the profession from very different backgrounds that did not include any form of medical study; quite often the town barber also served as the town surgeon.

As many communities could not support a barber-surgeon, a custom was developed so that some barber-surgeons traveled to perform needed surgeries. An additional benefit of itinerant surgeons (a benefit to the surgeons, that is) was that they often slipped out of town quickly and were not necessarily around if a patient became worse.

A barber-surgeon or other self-proclaimed surgeon was usually quite capable of removing small growths, fixing hernias, or setting broken bones, but other types of surgery were considered very risky. Unless the person was suffering from a condition or an ailment that had few curative options (the agony of kidney stones or the greatly reduced vision of *cataract*s among them) then surgeons avoided any type of invasive surgical operation, fearing a poor outcome. While a surgeon's liability was limited by law in many cities, these laws did not necessarily prevent an angry family member from seeking revenge against the surgeon if a patient died.

Surgery was particularly challenging during this period because physicians and surgeons had very little understanding of human anatomy. In the early Middle Ages, it was viewed as unacceptable to conduct studies of human cadavers, which limited the healers' ability to learn more. As a result, while some procedures were successful, surgery was a high-risk undertaking. This awareness frequently meant that physicians were reluctant to perform certain types of surgery.

This chapter will examine who became surgeons, how they were trained, what types of tools they used to perform which types of surgeries, and how the profession began to change by the latter part of the Middle Ages.

WHO BECAME SURGEONS AND HOW THEY WERE EDUCATED

The backgrounds of surgeons during this time varied greatly. Early in the Middle Ages, people frequently turned to clerics for help with medical issues, but clerics were generally forbidden from practicing surgery because they were not to "spill blood." Other scholars who trained as medical doctors considered surgery to be a "trade" and therefore, beneath them. If an injury or an illness required some type of cutting—or even the halting of bleeding—patients had to look elsewhere for help.

Townspeople turned to anyone who might be able to help them if they were wounded or suffered an injury. Because minor cuts often resulted when a barber was giving a close shave, all barbers kept bandages and *styptics* on hand and knew how to stop bleeding; when someone was hurt, the barber was a logical person to visit. Barbers also understood the locations of major *veins* and arteries because of their experience.

Small communities that could not support a barber-surgeon had to rely on the local healer, and his or her care was supplemented by traveling surgeons who occasionally stopped to visit the community. Some itinerant surgeons specialized in a particular type of surgery, such as "cutting for stone" or "couching for cataract." There were also traveling tooth-pullers. Those who came into town to see the ailing insisted on payment at the time of surgery so there was little incentive to remain until the fate of the patient was known. Because surgical knowledge was so primitive, there

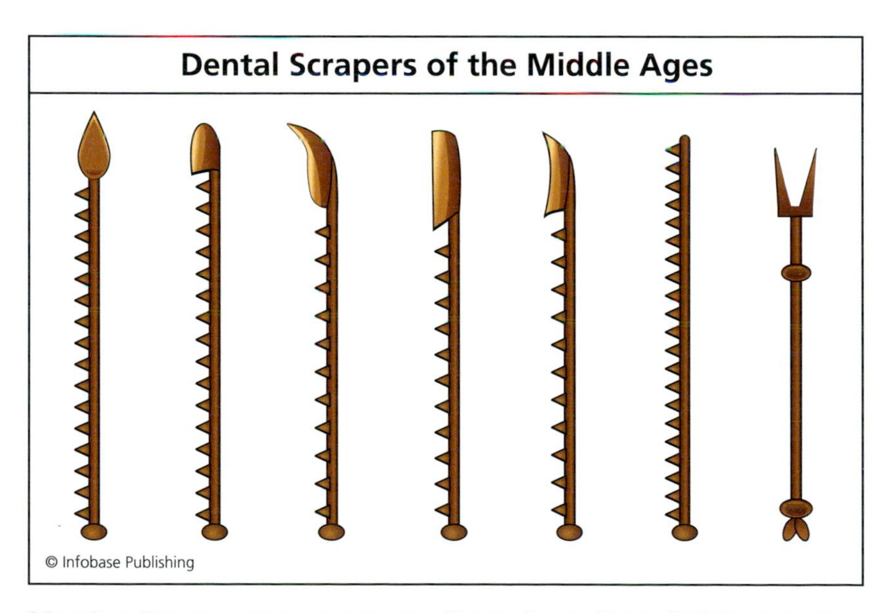

Mouth pain was common among all people, and one Arabian surgeon, Abu al-Qasim (Abulcasis; 936–1013), realized that the tartar that built up on teeth was not helpful to good tooth health. These are some of the dental scrapers he created to clean teeth.

were no guarantees. Many of the patients may not have recovered from the procedure itself, or even when a surgical procedure (such as a tooth extraction) was successful, there was little that could be done if an infection set in.

Education and Training

In addition to special training to learn surgery, today's surgeons have had to train as medical doctors; in medieval times, surgeons rarely studied any other form of medicine. The majority of surgeons learned their art as apprentices—first observing others and then practicing under the supervision of those with more experience.

By the 13th century a good number of barbers specialized in being barber-surgeons and no longer cut hair, focusing solely on surgery. As a group they organized into a *guild* and implemented apprenticeships for those whom they permitted to undertake study to become barber-surgeons. By this time there were a few physicians who practiced surgery, and they belonged to a different guild.

Professional guilds came to be a prominent part of village and city life during the Middle Ages. The members maintained control over the qualifications for membership, and they outlined training procedures, among other things. Connections to someone within the guild were important for those who wanted to gain access to a particular profession. A few women were permitted to become surgeons because they were married to or daughter of someone who was already a practicing guild member and was willing to train them. Women were believed to have impressive fine motor skills, so they were thought particularly well suited for surgery.

Later on when dissection became acceptable as a learning tool, surgeons began to be employed to participate in the lectures given by physicians in the universities. While the physician discussed a particular body part, the surgeon pointed it out—and enhanced what the physician talked about—through dissection of a cadaver. There were still many misunderstandings about the human body, so the accuracy of both the discussion and the dissection was not always reliable. Toward the end of the Middle Ages, surgery began to be taught in a few of the university programs, though the practice of surgery was still held in lower regard than the regular practice of medicine.

By the 14th century Italy began to employ surgeons for autopsies. Both physicians and surgeons were consulted in instances of suspicious deaths or when the cause or nature of the death or injury was in dispute.

SURGICAL TOOLS AND OPERATING CONDITIONS

Surgical tools remained unchanged throughout the period, and most of them were quite primitive. Because surgery was neither respected nor was it taught in the universities, there was little interest in advancing the following types of equipment used:

- Hacksaws were the basic instrument for limb removal.
- Mallets and chisels were used for *trephining* (skull surgery).
- Iron rods were employed for cautery (a process that destroys tissue through burning or scarring).
- Various sizes of knives and scalpels were used for different types of cutting.
- Forceps, tweezers, and tongs were used for the removal of foreign matter, holding open incisions, and repositioning tissue as necessary.
- Needles provided surgeons with the ability to close wounds through suturing.
- Small tubes, cannulae, were used for draining blood and other fluids from around internal wounds. One 13th-century Italian physician thought to insert a cannula into an esophagus of a patient to serve as a feeding tube.

These tools were primitive but reasonably effective in the hands of someone skilled. Unfortunately, the medical community knew very little about the importance of a sterile environment for operating,

This style of tool was used to separate the tissue for further surgical exploration.

and while wine was often used as an antiseptic, any cleansing that occurred was likely not thorough. If people were wealthy, surgeons had access to intricately decorated surgical tools to use on these patients, but the results were often worse than when simple tools were used, probably because of the inability of the decorated tools to be cleaned well.

When surgery was undertaken, the likelihood of internal or excessive bleeding and infection were just two of the problems that surgeons encountered. Medieval surgeons knew little about halting internal bleeding, and the science of *transfusions* was unknown at that time. Infection and death often resulted in the aftermath of surgery.

Relief from surgery-related pain was limited to alcohol or some type of sleep-inducing plant mixed into wine, neither of which were as effective or as controlled as today's anesthetics. The patient usually drank a prepared mixture, or, sometimes, the liquid was absorbed into a cloth that was held over a patient's nose or mouth. While these drugs offered some relief to a patient, the science of their use was still very experimental, and accidental poisonings or overdoses were common. Patients themselves sometimes solved the pain issue by passing out from the agony of their condition or from fear, and in most cases, surgeons had an easier time working on an unconscious patient, though this still didn't assure a positive outcome. A patient who was still conscious generally had to be restrained; illustrations from the time depict various ways patients were held or tied down so that surgery could take place.

TYPES OF SURGERY PERFORMED

When people sought the help of a surgeon, the patient was generally in need of direct hands-on care quickly. While a physician might have time to evaluate humoral balance and take into consideration the alignment of the planets before deciding upon a treatment, the surgeon generally needed to make a rapid evaluation and work quickly since blood loss had to be minimized. The elapsed time of the surgery needed to be brief since there was no good system for deadening pain.

Common surgical procedures that were performed without too much difficulty at this time included lancing a boil, dressing an ulcer or sore, repairing hernias, removing nasal polyps, and excising hemorrhoids. As had healers before them, medieval surgeons also performed trephination to release pressure in the skull. Repositioning a dislocated joint (usually a shoulder) required the physician to use both his hands and a foot to push or yank the shoulder back in place. A similar type of brute force was used on the jaw if that was the joint that had popped out of place. To set a bone, the surgeon learned to realign the ends of the bones and secure them with splints and bandages. There was a great enough need for bone setting that a few surgeons in larger communities actually specialized in it. Bandages were usually linen material that was soaked in egg whites and then placed over any wound. As the egg whites dried, the bandages contracted and pulled the patient's skin together over the injury. Surgeons also learned that dressings often needed to be replaced a few times during the healing process.

For cuts and other open wounds, the first step was to stop the bleeding, which generally involved using pressure or some type of styptic substance to stop the blood flow before cleaning and bandaging the wound. If a patient suffered a skull fracture, the surgeon used tweezers and tongs to probe the opening to locate and extract any bone fragments. An effort was made to drain off fluids while cleaning the injury and then clean linen was inserted inside the opening in the skull to keep foreign matter out of the wound. Skeletal evidence reveals that people survived some very deep skull injuries because they were treated successfully enough to heal.

Medical texts of the day address some issues like treating tears or cuts to the lungs, bowels, or other organs, and there are descriptions of malignant growths and how and when these were removed. Surgeons often resisted undertaking these types of surgical challenges since they were so often fatal. While historians have found reference to a few successful cases where a perforated bowel was repaired or a *tumor* was excised, these were rare examples.

"Cutting for stone" (removal of kidney stones; see the sidebar "Cutting for Stone" on page 65) and "couching for cataract" were two commonly needed and regularly performed procedures. As had people before them, medieval surgeons performed cataract removal by inserting a sharp instrument through a patient's cornea to force the lens of the eye out of its capsule and down to the bottom of the eye. (Restraining a patient was an important part of this procedure so that the person would remain totally still.) Severely damaged limbs had to be amputated by a surgeon, but with these cases, the outcome could not be guaranteed. Blood loss and infection were serious possibilities.

Eye surgery *(Medicine Through Time; www.medicinethroughtime.co.uk)*

Skeletal evidence indicates that many people who underwent the more minor levels of surgery survived; the more complex the surgery, the less likely was survival. Most patients who underwent these treatments probably did so with little or no anesthetic or sedatives.

BLOODLETTING, CUPPING, AND CAUTERY

Routine health care during this time often involved bloodletting. Also known as phlebotomy or venesection, the process of intentional bleeding as a cure for various ills dated to the height of the Roman Empire and was used as part of a treatment for rebalancing of the humors. As in Roman days, a physician or healer was responsible for determining how often and what part of the body should be bled.

To modern practitioners, bloodletting would be viewed as needless and quite possibly dangerous, so to keep this treatment in perspective, it is important to note that as early as the 12th century, medical texts stressed that only a moderate amount of blood should be taken. They also understood that pregnant women, small children, the very old, and the very weak should not be bled at all. Records from monasteries indicate that the monks knew that bleeding could be debilitating. Though payment records indicate that barber-surgeons visited monasteries regularly and performed bloodletting as part of routine health care (some monasteries record monthly bloodletting visits; at other locations, the visits were every three months), but the records note that monks were to be given extra meat to eat, time to relax, and less arduous chores during the first day or so after bloodletting.

Bloodletting

The actual bloodletting process involved making a small incision, often performed by a surgeon or barber-surgeon, to drain out some of the person's blood. The location of the bloodletting was determined by the nature of the illness as well as guidance obtained by the physician through interpretation of an astronomical calendar. Certain signs of the zodiac governed specific parts of the body, and it was believed that the health of the person would be affected by the position of the constellations. Illustrations usually showed the blood being caught in a small bowl so that it could be measured and examined. Leeches were used sometimes as they were effective at drawing out the blood, but because the quantity of blood removed could not be judged, they were less popular with medical people.

Many medieval people survived multiple bleedings, so the barber-surgeons must have become relatively skilled at bloodletting. Choice of vein, the amount of blood drained, and the method for closing the wound so that it could heal without becoming infected would all have been important in assuring the patient's well-being. Court records show that most problems that resulted from bleeding, such as accidental severing of nerves, the tendons, or the major arteries, were inflicted by empirics, not by trained individuals.

While nobles, clerics, and royalty may have been bled regularly, the majority of the population would not have access to this "state-of-the-art" health care. The process would have been costly, and people would have to travel to towns where there were barber-surgeons. As a result, the majority of the population would certainly not have been bled regularly. And as with rather

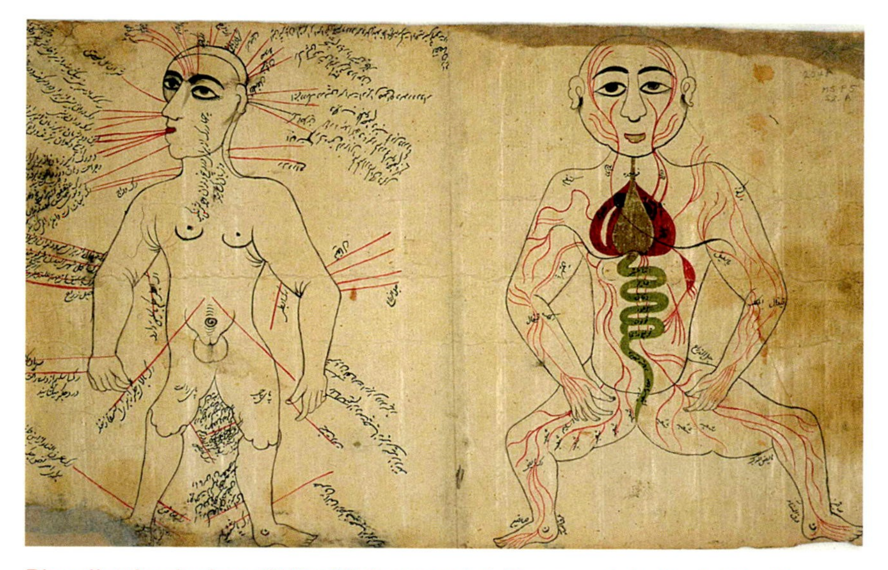

Bloodletting by Isma'il ibn Muhammad al-Husayn al-Jurjani *(The Treasure of Khvarazm Shah)*

unpleasant procedures today, such as dental visits, some people probably feared it and therefore avoided it.

Cupping

Cupping was another method used for draining blood in some illnesses. Surgeons made a number of small, shallow slashes in the patient's flesh at a location designated as significant to a particular illness. A metal cup was heated over an open flame, and then the hot cup was pressed against the scarified surface of the skin. The heat and the pressure created a slight vacuum and served to draw blood out of the cuts. Though this process involved both cutting and a very hot cup, it was considered less hazardous and less painful than bloodletting. As a result, it was often the first treatment of choice for the elderly as well as women and children.

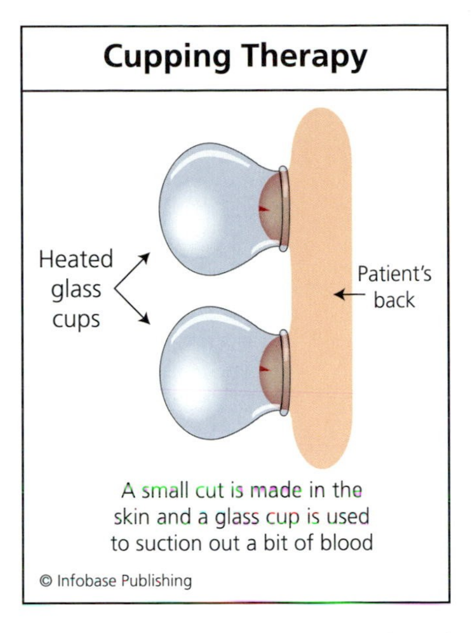

When a glass cup was heated and then pressed against the skin, it drew blood to that area. This was thought to be helpful to healing.

Cautery

Cautery is a process using either a chemical agent or a heated instrument to burn out diseased or severely damaged tissue along with any infection. Once the diseased tissue is removed, the heat helps seal the wound. During medieval times, cautery was used preventively, as well as to balance the humors. (The process was used by the Romans as early as the second century C.E., and the Arabs used it frequently as well.) By applying a red-hot tool at specific points for certain symptoms (much the way acupuncture and acupuncture points are used), it was believed the physician could adjust the humors and cure ailments ranging from headaches and joint pain to respiratory problems. Cautery would have been very painful, and the illustrations of such procedures often show bowls or cups nearby that probably contained painkilling or sleep-inducing solutions.

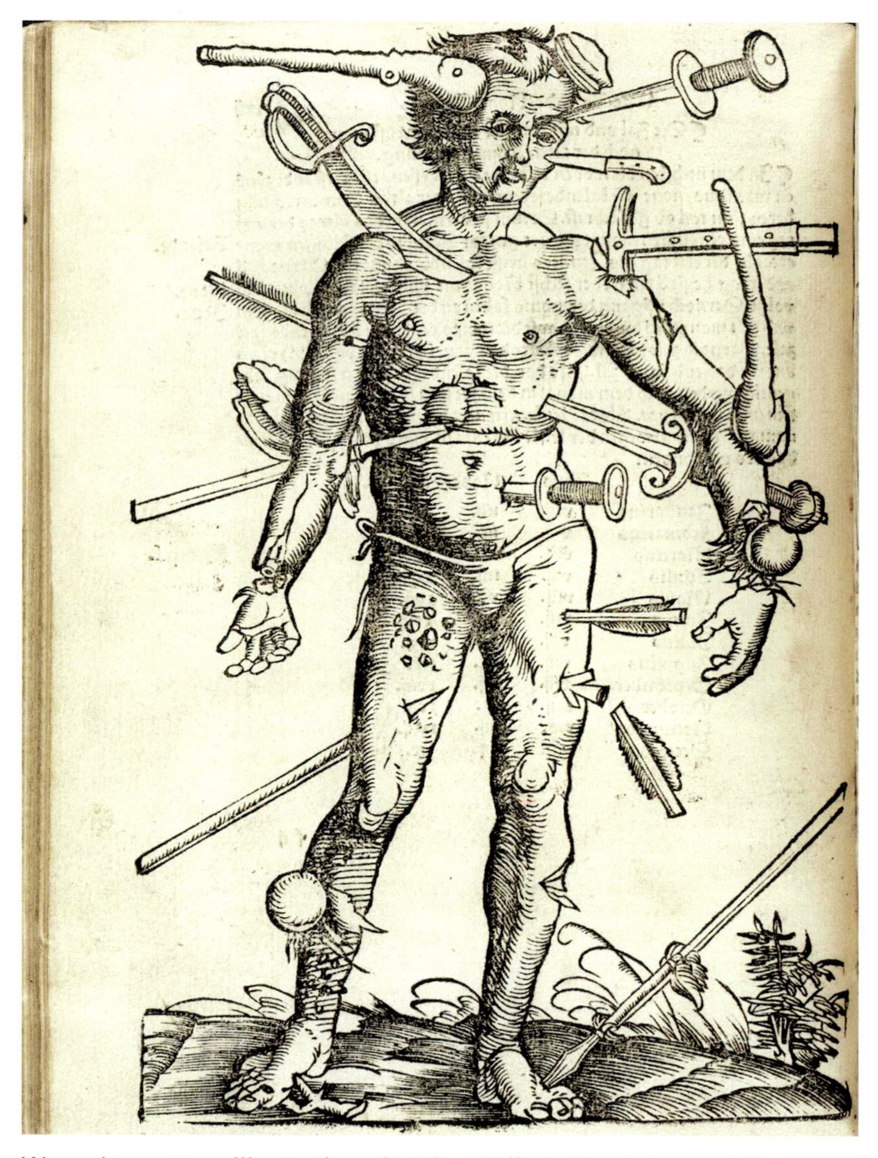

Wound man, an illustration that is attributed to surgeon Hans von Gersdorff from his *Feldtbüch der Wundartzney* (Fieldbook of wound surgery; Strasbourg, 1528) *(National Library of Medicine)*

CUTTING FOR STONE

One text from medieval times describes the procedure for the removal of a bladder stone, and David Lindberg quotes from it in his book *The Beginnings of Western Science:* "If there is a stone in the bladder make sure of it as follows: have a strong person sit on a bench, his feet on a stool; the patient sits on his lap, legs bound to his neck with a bandage, or steadied on the shoulders of the assistants. The physician stands before the patient and inserts two fingers of his right hand into the anus, pressing with his left fist over the patient's pubes. With his fingers engaging the bladder from above, let him work over all of it. If he finds a hard, firm pellet it is a stone in the bladder . . . If you want to extract the stone, precede it with light diet and *fasting* for two days beforehand. On the third day, locate the stone, bring it to the neck of the bladder; there, at the entrance with two fingers above the anus incise lengthwise with an instrument and extract the stone."

While cautery sounds crude, a form of it is actually used in modern microsurgery where surgeons use lasers to burn and seal blood vessels shut. Today patients have the benefit of anesthesia, a luxury that was not available to medieval patients.

The detail provided in this sidebar gives a graphic understanding of how stones were removed from the bladder during the Middle Ages. Though some patients might have been given wine beforehand, the procedure would have taken place without any other form of anesthesia.

BATTLEFIELD WOUNDS

Then as now, battlefield injuries were often fatal, and when not fatal, they were not easy to treat. Blood loss was always a problem, and injuries ranged from broken bones and head injuries to wounds from lances or knives, as well as damage from embedded arrows. Barber-surgeons were hired to accompany armies and treat wounds of the regular soldiers. A nobleman might retain both a physician and a surgeon to accompany him into battle. That way he was guaranteed to receive care by someone he knew and trusted.

Barber-surgeons' on-the-go surgical opportunities certainly increased their knowledge of human anatomy and helped to refine their skills. Blood loss was the first battlefield issue that needed to be addressed, and surgeons knew to use pressure, affix tourniquets or ligatures, and cauterize to slow blood loss. Abdominal puncture wounds and arrows that lodged into various parts of the body were the types of soft tissue wounds that had to be managed. Just as in war today, no sooner did they perfect a method for removing some type of weapon (usually an arrow), than the enemy developed a new type of arrow point that was more damaging and more difficult to remove. The risk of removing arrowheads or bone fragments from a nonfatal wound was high because of the very likely threat of infection.

Skeletons of soldiers from the time period reveal that other types of battlefield injuries included broken limbs and skull fractures. As described previously, a skull fracture required careful removal of bone fragments to avoid damaging the brain. Skeletal evidence shows that at least some of these soldiers got better.

Wounds frequently needed to be stitched up, and barber-surgeons carried with them needles and linen or silk thread. Texts indicate that they used wine to clean wounds when they could, and the alcohol in wine might have worked well enough as an astringent, but there were still many occasions when the wounds became infected.

The continued belief in the value of pus oozing from a wound was a very harmful idea of the time that was a carryover from Galen. The thinking of the day was that it was best to keep a wound partially open and moist with unguents so that the wound would produce pus. This theory seems to have been based on a misunderstanding of the healing process. Wounds that are healing properly will sometimes exude pus but practitioners of the "laudable pus" theory believed that wounds must always produce pus as part of proper healing. By the end of the Middle Ages, dry healing finally became acceptable. The new practice was much more successful. Wounds were usually completely cleaned, sealed, and dressed, and dry bandages were used to keep out contamination. The exceptions to this new practice concerned deep wounds or wounds to the abdomen that were sometimes kept partially open or fitted with cannulae to drain off fluids.

PROMINENT SURGEONS

During the 13th century, surgery became part of the curriculum at some universities, so some physicians began training to practice surgery, though it was still considered less prestigious than other types of healing. Lanfranc of Milan (1250–1306), a surgeon who was also known as Guido Lanfranchi or Lanfranco, was a native of Milan who became a popular professor of surgery at the Collège de St. Côme in France. The dean of the faculty requested that Lanfranc write about what he knew, and Lanfranc put on paper his knowledge of anatomy, embryology, ulcers, fistulae, fractures, and dislocated joints, as well as some nonsurgical subjects such as herbal medicines; the result became a major work on medicine. Though it was written in 1296 and was one of the few books that outlined the medical thinking of the day, it was not published until 1490. It became quite popular at that time and it was then translated into multiple languages and reprinted several times.

Shortly after Lanfranc, Henri de Mondeville (1269–1320), contributed to surgical knowledge. De Mondeville served as a military surgeon to the French royal family and he pushed for changes in wound care. Since the time of Hippocrates, healers had believed that wounds needed to create pus to draw poison from the body. De Mondeville recommended closing the wound before pus had time to form, and he is thought to have been first to prove that pus was not necessary in order for a wound to heal properly.

Guy de Chauliac's Ballista

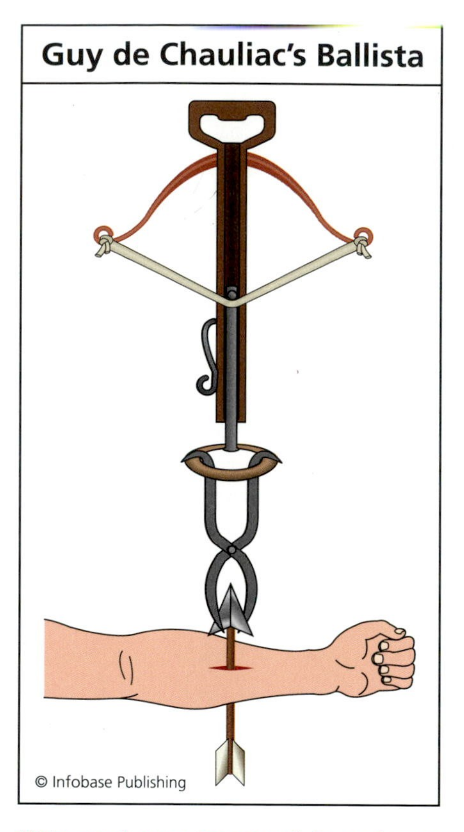

This tool was invented in order to ease the removal of arrows.

Guy de Chauliac (1298–1368), a disciple of de Mondeville, was a well-respected surgeon of the 14th century who is now known as the "father of surgery" because his writings were followed by so many and for so long. He was born in Chaulhac, Lozère, France, and was the physician for Pope Clement VI and two successors. He spent most of his life in Avignon, France, where he survived an infection of the black plague. After the *pandemic,* he wrote the medical reference about surgery, his *Chirurgia magna* (1363). He relied heavily on Galen.

The most prominent British surgeon of the time was John of Arderne (1307–92) who served John of Gaunt in the Hundred Years' War. He also is considered a father of surgery, both for his honorable conduct as well as the fact that he devised helpful solutions that made people feel better. One of his contributions to the field of surgery was for a repair of a fistula, which was a very common ailment suffered by knights who spent long hours on horseback. The affliction involved a large painful lump that appeared near the base of the spine. If untreated, it became an abscess and could rupture. Arderne devised a method for cutting the lump out and making a repair that let the knight go back to battle relatively quickly. He also developed an ointment that was helpful with arrow wounds.

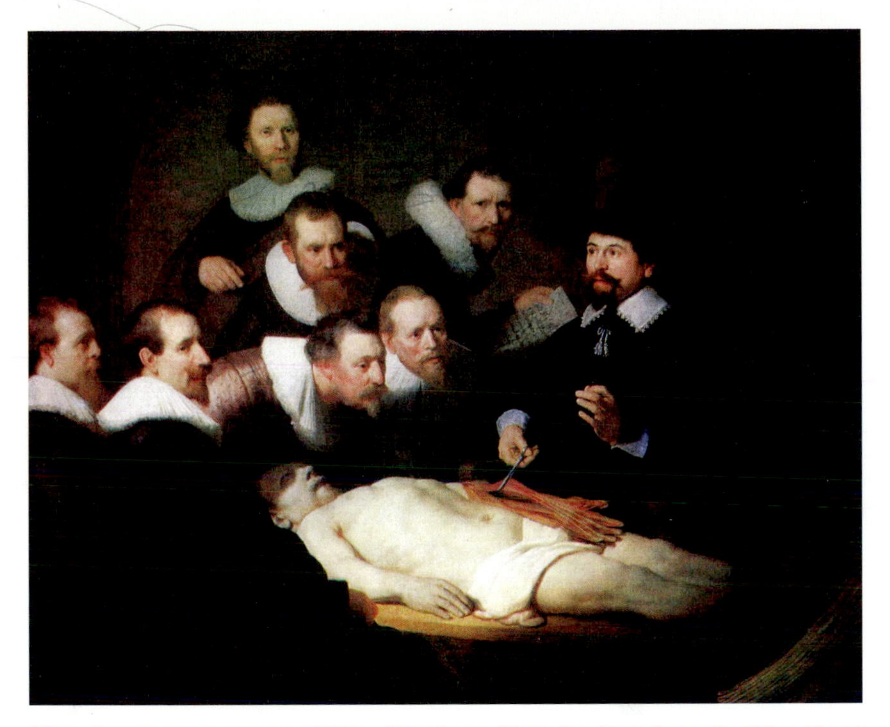

The Anatomy Lesson of Dr. Nicolaes Tulp by Rembrandt *(The Yorck Project)*

Arderne was also admired for his approach to fee setting. While he was happy to charge the rich for his services, he believed that poor men should be seen free of charge. By 1370 he was a member of the Guild of Surgeons, and by the end of his long life, he achieved the title of master surgeon.

CONCLUSION

There was a great need for surgical skill throughout the Middle Ages, but because of the high risk involved in surgery, physicians, whose reputations were created or destroyed by patient outcomes, refused to undertake most surgical procedures. This left the profession in the hands of barber-surgeons and other healers who learned the techniques through apprenticeships. While minor procedures could be performed with a likelihood of success, any invasive surgery ran the risk of resulting in heavy bleeding, infection, and death.

Guilds became important in limiting who trained and practiced surgery, and not until the end of the medieval period did a few universities begin to offer surgical training. There still was little knowledge of the workings of the human body, so even when training took place in a university setting, it did not necessarily improve the outcome.

Women Practitioners and What Was Known about Women's Health

Medieval healers were people with the time, availability, and interest in making people better, and there were few who were better suited for that role than the women of a community. The people of the time knew enough about disease to know that when someone was ill, this increased the risk that the illness might spread to others. In a feudal society everyone needed to participate, so when one—or many—fell sick this hurt the community's ability to survive. This provided women with the motivation to tend to the sick, and they gained invaluable on-the-job experience about the best ways to maintain good health.

Women were also vital to the healing process because men were prohibited from examining women. If a woman suffered from something serious, such as heavy bleeding or another type of dire medical situation, a physician might then be consulted. However, the female healer served as the intermediary since direct examination of a female by a male was not allowed.

Though women were primarily involved with family and female-related health issues, a few ventured into other fields. In

Naples, Italy, between the years 1273 and 1410, 24 women were listed as surgeons. (Women who served as surgeons were actually more common earlier in medieval times, as the rulings that barred women from practicing medicine and performing surgery became more restrictive later on.) In Frankfurt, Germany, between 1387 and 1497, 15 women were not identified as midwives but were specified as practicing medicine, so they must have worked as general healers or possibly surgeons.

During the 13th century, medicine began to be taught in a university setting, but with few exceptions, the study of medicine at universities was limited to men. Though only a few medical students graduated from each school per year, this altered the paradigm for the profession. Realistically, women were probably still very active as healers and they were vital as midwives. However, this new model of university-trained physicians, from which women were excluded, resulted in excluding women from any prestige involved in becoming a university-trained physician, as well as limiting their ability to achieve the increased earning power of those with more education.

Hildegard *(Wilfinger Hotels)*

The most significant women practitioners of this time period were Abbess Hildegard of Bingen, who wrote *Liber simplicis medicine* (Simple Book of Medicine) in about 1160; Margery Kempe, who is credited with passing down information about medicine of the time; and Trotula of Salerno, Italy, who became known for her work on women's illnesses, though much of Trotula's background is shrouded in mystery. Some scholars today think that

Trotula may have been a fictional figure, whose name was used to compile information on diseases that are common to females.

This chapter will introduce the women who contributed to advances in medical care, describe what was known about women's health at this time, and remind people that the fear of witchcraft prevented many women from practicing any form of healing.

ACCEPTANCE IN THE FIELD

When someone in a community was ill, women of all classes were the first line of defense. Unable to afford any other type of healer, peasant families turned to the women among them who could help a sick family member. Wealthy women were not only expected to tend to health issues in their own family but they were often sought out to help with servant families as well. In *Le Managier de Paris,* a book written in the 14th century, a wife is instructed to drop everything if a servant falls ill. Noble women sometimes provided what was considered a slightly higher level of care because some had enough formal education that they had learned Latin. This permitted them to look for additional information in medical texts that had been preserved from earlier times and translated into Latin. Some communities relied on "wise women" who were called in if additional help beyond the family was needed. Many of these women gained knowledge by working as apprentices to surgeons.

Women also served as nurses in monasteries. Some monasteries ministered only to the health of monks or priests yet at other times the facilities opened their doors to all comers. Women who served in hospitals took care of all the housekeeping duties of the hospital, including tending to the garden where much of the food and medicine was grown. They prepared meals, did the laundry, and kept the rooms clean. Many of these women were part of a religious order, and were referred to as "sisters" because of this. A few of the women served at a higher level and operated as adjunct practitioners to physicians. A small number also served as apothecaries, specializing in the mixing of various herbs.

This 15th-century painting depicts Hildegard (kneeling) as a child *(Karey Swan)*

Professional guilds that regulated who could practice medicine and how they should be trained as well as the licensing of medical practitioners developed as a way to control who practiced medicine in each community, and women must have been very necessary to many communities as a good number were given licenses to care for patients. A guild in Florence permitted a few women to join, and in 1276 Sister Ann of York (England) qualified to practice medicine. Most who did so were related to a licensed physician or guild member, and they took over a practice left by a brother, father, or spouse.

If a woman had a family connection to someone willing to train her, then she might have become a surgeon, a job for which women were thought to be well suited because it was believed that women had fine motor skills. In about 1250, a woman identified as "Katherine" became a surgeon in London; she was the daughter of one surgeon and the sister of another. Agnes Woodcock, a British citizen, may have been among the last of women during medieval times who received any type of training in surgery. Agnes apprenticed under surgeon Nicholas Bradmore, who thought enough of her work that he specified in his will that she should receive a special belt and money when he died (1417). By 1380 some guilds required an oath to be taken by surgeons, vowing to provide worthy oversight of the profession, and there is evidence that women were permitted to take this oath so a few women must have been permitted to continue practicing at this time. The information on these women is sparse, so they may not have been very public in offering their services.

By the 14th century, men were beginning to enter universities to train as physicians, and as the movement toward university training solidified, it created another obstacle that prevented women from practicing medicine. While a handful of women were permitted to study medicine at the University in Salerno, this was an isolated situation. Dame Trota, also known as Trotula, is included among this group of women, but as historians have looked more deeply into her past, some feel that Trotula was apocryphal. Medical historian Roy Porter feels that Trotula may have been a male doctor who took on the persona of a female because of his intention to write about femalecentric topics such as conception, pregnancy, childbirth, and motherhood.

Though the number of people who actually graduated from universities with medical degrees remained very small, the movement to prohibit women from practicing medicine expanded. In Britain in 1421 English physician Gilbert Kymer and some colleagues submitted a petition to the British Parliament to ban women from working as doctors. Over time the working environment became more and more restrictive, and women were confined to nursing, midwifery, and home care.

CHILDBIRTH AND FEMALE HEALTH ISSUES

The Romans had developed some helpful concepts regarding childbirth and gynecological care, but these beliefs were not destined to be part of medieval health care. Though Soranus of Ephesus (a Roman from the second century C.E.) wrote knowledgeably about the diseases of women and infants, and he devised a method to

MEDIEVAL UNDERSTANDING OF WOMEN'S HEALTH

The people of the Middle Ages had many misconceptions about women and health that were part of the mythology that surrounded the female of the species. No male physician ever had direct contact with a female patient, so it was easy for the lore to grow. For example, they believed that women were constitutionally weaker and more prone to illnesses than men. As in earlier times, medieval practitioners attributed many female illnesses to a "wandering uterus" and the "cure" was to have a woman inhale noxious fumes in order to get the uterus back into the proper position. They realized that the menstrual blood must in some way provide nourishment for new life, so in that regard it was seen as a vital part of fertility. But menstruation was also viewed as a sign of weakness, and during the menstrual period women were considered unstable.

Medieval understanding of conception, referred to as "generation," was a composite of teachings from the Bible, Aristotle, the Greek physician Soranus, and Galen. Theories on conception were seriously misguided. A few people believed that only the man's semen was necessary for conception to occur. However, if the couple failed to produce offspring, women were solely to blame. Baby boys were preferred, and medical texts and manuals often offered suggestions as to how to increase the likelihood of giving birth to a male. Some of the recommendations on this topic included avoiding distressing sights during pregnancy.

The church forbade birth control, but social scientists who have analyzed population figures feel as though many citizens used some method of birth control or the population figures would be higher. This finding is validated by the fact that various methods of contraception are described in two books, the *Trotula* and *The Secrets of Women,* so there was information available at this time on how to prevent pregnancy.

turn an infant in the womb in order to facilitate a safer birth, these theories were forgotten during the Middle Ages.

Other than Hildegard's books (see page 79, "Hildegard of Bingen (1098–1179): Respected Healer"), there were very few books about medical treatment of women. Peter of Spain (who was to become Pope John XXI) wrote *The Treasury of the Poor,* and it contained more than 100 prescriptions for aphrodisiacs, fertility, and contraceptives. Otherwise, women generally had to turn to folk remedies for cures for anything from yeast or urinary infections to headaches and chest congestion. A few texts noted remedies for discomforts specifically suffered during pregnancy such as swollen feet or painful breasts.

Whether Trotula was a pen name for a male doctor or actually a university-trained female physician as originally claimed, Trotula wrote a health manual that was not particularly helpful. Though the author correctly indicated that the placenta needs to be expelled after birth, the methods the book recommends to do so are nothing short of bizarre. Among them were these: burning the bones of salted fish, horses' hooves, or the dung of a cat or lamb so that the smoke "fumigates the woman from below."

Childbirth was very dangerous; both mother and baby were at risk. Hemorrhaging, a prolapsed uterus, and a retained placenta were all very serious complications of childbirth. There were few effective solutions. If a woman died in childbirth, then a surgeon or male physician was summoned to attempt to save the baby by performing a cesarean section on the deceased mother. They did not yet have the medical knowledge to permit a woman to survive a cesarean, so it was only undertaken when the mother died.

Women generally gave birth in a "confinement room" using a birthing chair, which was designed with a V-opening so that midwives could easily check on how the birth was proceeding. The chair would have been a positive invention, since it put women in a more or less upright position, encouraging gravity to help the birth along. If no birthing chair was available then a woman probably crouched or was propped up. Oils were applied to reduce tearing, and stitches were used to repair the area if the vaginal opening ripped. Midwives, neighbors, friends, or maids oversaw

The birthing chair permitted a woman to remain upright while giving birth and thereby would permit gravity to ease the process slightly.

childbirth. Men were not permitted to attend to women in labor, though male physicians were sometimes consulted if there were difficult circumstances.

After a successful birth, family and friends customarily held a special postpartum celebration and provided the mother with new clothing to wear. Well-to-do women were given herbal remedies that successfully suppressed lactation, and their infants were given to wet nurses to be fed and cared for.

The risks to children during childhood were almost as great as the risks to mothers during childbirth. Birth defects, infectious diseases, and injuries during birth as well as during toddlerhood were common, and children frequently died.

MIDWIVES

Even after the 14th century when university training became important within the medical profession and women started being edged out, women were still valued for their work as midwives. Like most medical practices of the time, midwifery alternated from the practical (helping the woman give birth) to the superstitious. Among the superstitions was the belief that the mother would have an easier labor if the midwife opened the doors and drawers and cabinets of the household, took out the stoppers of every bottle, jar, and jug, and removed all the hairpins from the future mother's hair.

While few texts on *gynecology* were to be found during the Middle Ages, one that survived from Roman times noted the qualities of a good midwife. It specified that she should be physically robust, disciplined, sympathetic, sober, discrete, and calm and possess a good memory and knowledge of medicine. Ideally, she was to work under a physician, but the dearth of medical practitioners meant that midwives frequently had to function on their own.

Many midwives were the wives or daughters of local physicians who learned their trade from a more experienced midwife. Most had gone through pregnancy and given birth themselves, and actual medical knowledge varied greatly. Few had formal education but often they were more knowledgeable than the doctors because of all they learned through experience.

By the 14th century, cities in France and Germany required midwives to be examined and licensed before practicing. Clergymen or physicians often conducted exams but in one German city an elite group of women carried out the licensing procedure. A few German towns were progressive enough to employ midwives to help with pregnant women regardless of their ability to pay.

HILDEGARD OF BINGEN (1098–1179): RESPECTED HEALER

During the Middle Ages, one woman, Hildegard of Bingen, rose to prominence as a healer and is well remembered because she wrote extensively about medical conditions and cures. Hildegard was the tenth child born into a family of free nobles who lived near what is now Frankfurt, Germany; it was a custom of the time for families to "tithe" their tenth child to the church. Ten children were a lot to clothe and feed, and so this helped lighten a large family's burden while providing the church with an additional person to work for the church. Hildegard was placed in the care of Jutta, the sister of Count Meinhard of Sponheim, where Hildegard learned the ways of religion and life. Jutta died in 1136, and Hildegard became the leader of the female religious community over which Jutta had presided. Eventually she became abbess of a convent at the monastery of Rupertsberg.

From childhood on, Hildegard suffered visions that debilitated her. Though she kept these episodes a secret for a long time, she eventually wrote of them, attributing the visions to religious causes and believing that what she saw represented the word of God. She then interpreted her visions as to what to do to cure others. (Two of the books she wrote were specifically about these visions and her interpretation of their divine meaning as it applied to healing.) Hildegard described her visions as intense light followed by "extinguished stars," after which she was physically spent.

British-born Oliver Sacks, now a U.S.-based neurologist, has studied Hildegard's symptoms and feels that her visions were actually the aura that precedes many people's migraines. He viewed her postvision symptoms as those of classic migraines, where the sufferer often suffers nausea, a temporary feeling of paralysis, and blindness, only rebounding later on. Writes Sacks in his book *Migraine*: "Among the strangest and most intense symptoms of migraine aura, and the most difficult of description and analysis, are the occurrences of feelings of sudden familiarity and certitude . . . or its opposite. Such states are experienced, momentarily and occasionally, by everyone; their occurrence in migraine auras is marked by their overwhelming intensity and relatively long duration." This explanation gives credence to Hildegard's certainty that her visions were "real." Hildegard's ability to turn what was a debilitating illness into words from God says a lot about her ability to turn an unfortunate condition into a way of helping others.

Her Works

Hildegard wrote two major books of medical writings after she received what she felt were her "divine revelations" about the causes and cures of many diseases. *Causae et curae* (Holistic Healing) and *Physica of liber subtilitatum diversarum naturatum creatarum* (The Book of Simple Medicine). She based the science in her books on the Greek cosmology of the four elements—fire, air, water, and earth balanced by heat, dryness, moisture, and cold. *The Book of Simple Medicine* was encyclopedic in its scope, and while much of it was possibly helpful, it also included magic formulas.

As her work expanded, Hildegard addressed many aspects of healing; she wrote of using plants, trees, and stones medicinally.

Her advice ranged from overcoming a simple cough to what to do about leprosy. The ingredients used included easily found herbs as well as items such as "lion's heart" (probably the local name for a

A painting of Hildegard from ca. 1165 *(Christian Rohr)*

specific plant). Her recommendations included cannabis to relieve headaches, nutmeg as a purifier of the senses, and rose leaves to clear sight. She wrote that plants and herbs were God's gifts. If they did not make someone better, then God did not intend the person to get well.

Hildegard also wrote about skin disease, including leprosy, scabies, lice, insect bites, burns, and various allergic responses, and she noted appropriate therapies for each. Many of her remedies were also intended to work through the skin, delivered through baths, sauna, or various types of ointments.

Her books offered practical remedies that were consistent with her religious philosophy of moderate living as well as taking responsibility for one's own health, and she believed that taking pleasure in walking, mountain climbing, swimming, rowing, fishing, gardening, and music and painting all helped to improve health.

After writing *Physica* (Natural Science) she achieved great renown for laying the foundation of botanic studies in northern Europe. Her efforts to outline the properties and uses of so many plants and items from nature were influential to botanists.

At a time when few women were in positions of authority, Hildegard garnered respect for her work, and bishops, popes, and kings consulted her. She left a legacy of natural healing knowledge, and today those interested in natural healing have revived it. The "plant" section of one of her books, *Physica,* has been recently reissued as a manual on using herbs and plants for healing. (Most of the items written about are still available in today's health food stores.)

Hildegard also composed music and her musical works are still known today.

MARGERY KEMPE (1393–CA. AFTER 1438): PRACTICAL MEDICINE V. SPIRITUAL CARE

Margery Kempe (1393–ca. after 1438) is said to have written *The Book of Margery Kempe,* a work that depicts the relationship between practical medicine and spiritual care during the Middle Ages. After Kempe gave birth to her first child (the first of 14), she began having visions. Some viewed her as mad; others thought she was a mystic. Medical historians note that her strange behavior may have been the result of puerperal fever. (This illness occurred in women after childbirth as a result of unsterile obstetrical procedures.) She soon sold her business, a brewery in King's Lynn, England, and devoted herself to religion, making pilgrimages to Jerusalem, Rome, and Spain. She eventually wrote about her spiritual experiences. Though this book is considered the first autobiography, it may have been written by a family member as there is no evidence that Kempe herself could read or write.

The medical care she describes in the book is holistic and is depicted as involving a great deal of social involvement in both births and deaths. This is a perspective that is missing from the general health manuals that have survived from medieval times.

Some of Kempe's views were considered heretical to what people of the Reformation believed. Had a manuscript of the book not been found in Lancashire, England, in 1934 in the library collection of the Butler-Bowdon family, historians probably never would have benefited from Kempe's experiences. If other copies of the manuscript existed they were very likely destroyed by religious devotees who would have found them objectionable.

KEEPING WOMEN OUT OF MEDICINE

In Paris in 1322, five women were put on trial for practicing medicine without a license. The charges against one of the women specified that she "visited the sick, felt their pulses, examined their urine, and touched their limbs." Eight patients came forward to speak up for one of the women, Jacqueline Felicie de Almania, whom they said had cured them when others had failed. Despite this positive testimony, the judge still found all of the women guilty. Their punishment was excommunication from the church, a harsh punishment for the time. This meant that they were not permitted to attend church ever again, and the ruling noted that they would "go to Hell when they died." The judge concluded with this: "It is certain that a man qualified in medicine could cure the sick better than any woman."

Some of the rituals that accompanied the herbal remedies had pagan origins, and this led to an effort to identify possible witches within a community and bring them to trial for practicing witchcraft. A book, *Malleus maleficarum* (*The Witch Hammer*), written in 1486 by two Dominican friars, Jacob Sprenger and Heinrich Kramer, focused on their pursuit of women who were practicing witchcraft. Their writings warned against women who were healers and midwives who might have been practicing witchcraft and offered methods for seeking out and destroying them. Over time, 50,000 women were burned or tortured for being witches during medieval times.

CONCLUSION

During the Middle Ages women were important participants in the field of medicine. They were people to whom a community could turn for help, and their on-the-job experience coupled with their good instincts meant that many of these women excelled as healers. Though the trend toward university training for physicians lowered the status of women who were healers, in actual practice, women were still very much needed as the only resource other women could consult and for their ability and availability.

6

Public Health in the Middle Ages

The Middle Ages in Europe are generally portrayed as a period when filth of all types abounded. The streets were reputed to be filled with garbage and waste materials that households freely dumped out their doors or windows. Sewers existed but they were little more than open channels through which the sewage could flow, so towns were very likely smelly. The practice of bathing varied by region; some communities had well-used public baths, but elsewhere bathing was discouraged since some believed that warm water opened the pores and this led to ill health.

Though contemporary documents on this subject are not plentiful because of the low literacy rate, analysis of facts from the period led historians such as Paul B. Newman, an expert on medieval life, to the conclusion that the truth lies somewhere between the myth of dirty streets and the ideal of the level of cleanliness achieved by the Roman Empire. The population figures during the Middle Ages reveal steady growth until the 14th century when there was a severe drop in numbers because of the ferocity of the bubonic plague. The fact that towns and cities supported strong growth is evidence that the medieval towns had developed some methods for effective sanitation techniques or disease would have been more rampant. Cases of dysentery and similar illnesses seemed to surge around the periods when a medieval town was under siege, meaning that when a community was under stress, there must have been greater difficulty in establishing good conditions for public health.

Other evidence of the fact that communities paid attention to public health is found in the governmental laws of the time. Records show that there were fines for people who dumped waste out of their windows. There were regulations on the locations of cesspools and city laws against accumulating garbage in the street.

Because a population's health is directly affected by access to clean water, uncontaminated food, and an acceptable level of cleanliness in one's living environment, this chapter examines the delivery of water to various communities during the Middle Ages, the provisions for the removal of waste, the public laws enforcing community hygiene, and the popular views on the advisability of bathing.

THE LEGACY OF THE ROMANS

As the Roman Empire expanded into Europe, the Romans continued to construct the infrastructures they had always created in their communities. Big stone *aqueducts* were built to bring in fresh water, and sewers were dug to take away waste in the new areas settled by the Romans. However, not all of Europe benefited from what the Romans knew. The Empire had never expanded as far as Scandinavia or northern Britain, and in central and southern Britain as well as other peripheral areas of the Empire, the Roman settlements often did not feature the more advanced aspects of some of the Roman developments.

Though the Empire was crumbling by the Middle Ages, the infrastructure of many of the communities remained and was put to use by the new inhabitants of each area. Because the flow of both the water and the waste materials relied on the very simple principle of gravity, these structures were easy to maintain; residents were able to continue to use them with a minimal amount

WATER DELIVERY SYSTEMS

The medieval people had not yet conceived of ways to pump water to various destinations, so gravity-powered delivery systems prevailed, sometimes bringing water in from several miles away. Pipes were sometimes created from hollowed-out tree trunks, particularly from durable hardwoods such as elm. Workers created a long tube by boring through the length of the tree trunk, leaving one end slightly flared while the other end was tapered. This permitted the two ends to fit together and create pipes that could carry water for longer distances. The earliest pipes had to have the inner opening bored out by hand using large augers, but eventually water-powered boring machines were developed.

For metal pipes, lead was generally the substance of choice, using a method developed by Romans (*plumbum* is Latin for *lead*). Lead was relatively common; it was less likely to corrode than iron and because it has a low melting temperature (it had to be melted, flattened, and rolled into a tube), the substance was relatively easy to work with. Though many modern day communities still have lead pipes carrying water, the metal is no longer used because it is now known that the lead leaches into the water and can cause health problems. This is a recent discovery, and in earlier times lead was considered an excellent building material.

Eventually some of the medieval pipes contained features that are common in plumbing of today, including Y- and T-shaped joints, vents, and valves for controlling the water as it moved through the pipe system. Though the Greeks had created small pumps to help move water, medieval towns seemed to rely primarily on gravity for water movement. Though the pipes were relatively large, which would have

(continues)

(continued)

increased the water flow, the absence of pumps would have meant that the flow of water would have lacked pressure.

The systems were usually designed so that water could be slowed or held temporarily in "settling houses" along the way. This allowed the sediment to settle out before the water continued through the pipes and into a cistern. The early methods of water purification were very primitive. Metal grillwork was inserted at places where the pipes were joined, and this prevented logs, weeds, fish, and animal carcasses from flowing along with the water.

of repair work. As more time passed and these facilities fell into greater disrepair, the citizens of the Middle Ages did not always work to preserve them. In villages that were expanding, workers sometimes took the more easily available stones from the aqueducts and used them as building materials. As a result, this Roman legacy disappeared in many locations. While the medieval people continued to create their own sanitation facilities and waterworks, they never mastered the art of building on the grand scale that the Romans did.

CLEAN WATER

At the beginning of the Middle Ages, much of the population lived in rural areas with no public infrastructure. As a result, the two areas where large groups of people lived together—monasteries and castles—proved to be the locations where progress was made on water and waste management.

In monasteries, the water traveled through a piping system and was made available for use in the monastery's kitchen, the brewery, and the latrine. Hand basins with running water tended to be located near the infirmary and/or the dining hall as priority was placed on washing hands before eating as well as before caring for the sick. (Bacteria and the scientific reasons for handwashing were not yet known but medieval people had some understanding that cleanliness was a good idea.) Some monasteries had hand-washing fountains (sometimes called lavabos) where water flowed out of multiple outlets into a circular basin. Faucets were added to control the water flow, and sinks had drains that took the water out into pipes and carried it away. In some cases, the used water entered pipes that then flowed through trough areas with waste, so the sink water helped wash away filth.

Palaces and castles also required living space for a large population, and they were built with some of the same construction methods used for monasteries. (Castles mainly housed soldiers so they were simpler structures than palaces.) These buildings had to be located strategically—often on high ground—to provide protection against attack. Though water might be nearby, it was more difficult to pipe in fresh water because the water was often located in areas that were below the palace or castle. Digging a well was possible, but it was very hard work and took time, and cisterns could be used to collect rainwater; however, the only way to guarantee enough water generally involved bringing it in by the barrel or in large waterskin bags.

Towns and small communities were well located if they had a good water source nearby, preferably one that was at a higher elevation. This permitted gravity-driven water to be brought in by pipes or water channels. (Often the Roman aqueducts could be reused for delivering this water.) Rivers and streams near population centers often became contaminated by sewage and other garbage, so developing sanitation systems and regulations was important (see the following section, "Sanitation Systems"). To supplement the fresh water available in the area, towns arranged for delivery of immense casks of spring water by barge or cart. The water was then emptied into public cisterns to service the town fountains and pipe systems.

Even in towns with some type of water system, water was a fought-over commodity. Towns often grew more quickly than public works systems could be built and families competed with businesses, including brewers and butchers, for use of the water. Those "upstream" of the system frequently took too much water, leaving little or none for people down the line.

In acknowledgment of these issues, cities enacted regulations to control water usage. There were annual fees for water service and a restriction on how much water could be used, and there were penalties for anyone tapping into the water illegally.

To provide water for themselves and their families, people generally carried waterskin bags or buckets to a nearby river, spring, or public fountain to fill them with water. Water is heavy so this process was arduous, meaning that the water tended to be used very sparingly.

By the 13th century, a few private homes in prosperous cities sometimes had their own water supply piped into the home, and the inside water flow was controlled by a faucet. The pipes used were very narrow so the water delivered would not have been abundant in flow or quantity, but water brought into the home was a luxury that did not exist in most homes until the 19th century.

SANITATION SYSTEMS

As with water delivery, castles and monasteries were the first locations where relatively high population numbers drove a need to develop methods for disposing of waste. While a castle's elevated location complicated water delivery, it simplified waste removal. When possible, the planner-architects of the day preferred to place buildings where there was easy access for piping to reach to a long, straight drop, preferably over a river or moat where the water would take away the waste. One monastery was built on a coastline and the waste was flushed into an area where rising tides could carry the waste out to sea twice a day. The pipes or troughs carrying waste and kitchen garbage angled down and joined the main drainage shaft, with everything exiting from one opening that dumped into a waterway or the moat. This opening had to be covered with an iron grating to prevent people from entering the shaft and gaining access to the castle.

By the 12th century (perhaps earlier) some monasteries and castles created methods for flushing without the tides. They constructed large cisterns; by elevating them, the released water had enough pressure to carry away waste.

In towns, sometimes there was no way to use area waterways for disposal, so the townspeople dug *cesspits,* which were usually lined with stones or wooden planks pierced with small holes. The intent of the pit was to allow the liquid waste to leach slowly into the ground while the solids accumulated in the pit. Some latrines were directly over cesspits (as today's mountain or park outhouses are); others were dug a slight distance away and the waste was carried via water down a shaft or pipe into the pit. This permitted the pit to be covered more like an underground cave; this also helped diminish the odor from the waste. When waste was dumped into cesspits or ditches instead of flowing rivers where the waste could be carried away, *gongfermors* had to be hired to dig out and carry away the waste, work usually performed at night. This was an unpleasant job and financial records from medieval times reflect that these workers were well paid.

Towns needed to plan placement of cesspools carefully because wells could become contaminated as sewage from cesspools in the area sometimes seeped into the well water. Cesspit regulations became part of the zoning code in London as early as 1189. It was specified that the cesspits for private residences had to be at least 5.5 feet (1.7 m) inside property lines though stone-lined pits could be built within 2.5 feet (.8 m). People came up with creative ways to dispose of their own waste. One woman ran an illegal pipe to the street gutter to take away sewage, and another homeowner piped his waste into a neighbor's cellar.

In London, dumping into the Thames was forbidden, and it was specified that conduits were to be flushed now and then. While the river was still very dirty, these measures helped reduce waste and keep the river channels open. During the 14th century, one stream in London became so clogged with intentional runoff from area latrines (the method used for having waste carried away) that the city officials ordered that all latrines in the area should be torn down. While the water may have become cleaner during this time, latrines were obviously necessary, so later the community arrived at a compromise. The city permitted residents to rebuild latrines that emptied into the stream, but they had to pay a fee for using the stream. There was also an ordinance against dumping any other type of waste into the city waterways. Unfortunately, the increasing population in the area eventually led to an increase in pollution of the stream, and by the mid-15th century the city finally ordered the stream to be paved over.

Northern Italy also issued regulations for its rivers as did Paris for the Seine. In 1550 Parisian municipal officials successfully opposed a plan that would have permitted some upstream sewers to empty directly into the river.

Roman tradition dictated that collection of urine for commercial use could be resold to wool processors, and so there was often a systematic method for collecting and removing urine from urinals in public toilets. Because urine is rich in ammonia, it is good for removing oils from sheep's wool and this was an essential step in turning raw wool into good cloth.

Solid waste then and now was more problematic. They understood that human waste could not be used as a fertilizer component though animal waste could.

"Necessary Rooms"

Urinals in monasteries generally consisted of a trough in the floor along one wall, though a few had more elaborate systems with a basin around a central fountain to take away the waste. The latrines in monasteries consisted of long wooden planks, placed on wooden or stone supports, with appropriately sized holes placed at regular intervals. They were referred to as *necessarium* (Latin for "necessary room or place"). Archaeological evidence reveals that though they lacked toilet paper some monasteries used thin rags, possibly torn from old robes.

Toilets in palaces were referred to as *garderobes* (French for "clothes closet" but this was probably just a way of being discrete). Their design was essentially the same as the ones in monasteries with an emphasis on placing toilets in locations that simplified waste disposal. Generally toilets were designed to be used by multiple people at the same time, but occasionally there were private ones. Some palaces had walls that were six feet or more in thickness so occasionally small rooms with toilets were tucked within the very thick walls. Some backed up to a chimney, which would have provided warmth within the room as well as an opportunity to vent air into the chimney. Scarcity of flushing water meant that the toilets usually relied on gravity and were built more like outhouses with wooden seats over openings; however, a few were tied into the palace water system creating a basic method for flushing.

One archbishop created a separate tower for the garderobes that could be reached through passages on each floor of the building. Stone partitions divided each of the toilets from the others, and all of them backed up to a central shaft into which waste from all the toilets could empty. This circular layout combined with the narrow stone partitions between the toilets permitted several people to use the facilities at the same time but in privacy.

In the late Middle Ages, chamber pots and chairs with a chamber pot underneath may have been used in some palaces but based on the number of toilets they may not have been necessary.

For peasants and townspeople, there were a variety of solutions to waste disposal. On farms, outhouses were located near the home, and in towns private homes sometimes had indoor facilities or the people used a combination of outhouses and public toilets. Chamber pots and buckets were also used for waste.

THE PUBLIC BATHS

During the Middle Ages, bathing customs varied greatly by location, and they evolved over time. The Romans had viewed bathing as pleasurable; it often took place in mixed company, and women would get into the baths in lavish jewelry and headdresses. (Veils were draped over married women for reasons of decency.) Food was often served on floating tables and it was very much a party atmosphere. While some of this tradition continued among the upper classes in the early Middle Ages, the church took a dim view of this practice and felt that excessive attendance at the public baths was not a good idea. By the fifth century church leaders were making it clear that bathing should be restricted to what was necessary for cleanliness and health.

As bathing customs evolved, separation of the sexes was customary in many locations. Some public baths had men's hours and women's hours. Other baths were larger and offered two separate areas that were divided by grillwork so that the sexes did not intermingle. In communities where bathing involved mixed genders, men were dressed in something that covered their genitals and women wore linen shirts to the knees. Men and women donned head coverings for health reasons, and women tended to wear their jewelry. In less desirable neighborhoods, many people went to public baths naked so nothing could be stolen while they were bathing.

By the late 13th century the bathhouses in Paris employed criers to announce when the water was hot. If a community was well organized, the bakers were notified of the heating of the bath furnaces. That way the bakers could use the bath furnaces to bake their products after the water had been heated.

In northern and eastern Europe, steam bathing became popular. Rocks or stones were heated within a closed area, and then water was applied; this created steam. (During the 12th century, Hildegard wrote of German "vapor" baths, and this was likely what she was describing.) People came in to sit in the steam. Those who took steam baths were generally naked except for their hats. Beating oneself with bunches of leaves was customary to encourage circulation. Afterward, people either cleansed in cool water in tubs or plunged into a cool pond.

In the 13th century a few communities had laws that required that every citizen visit a public bath weekly. If they failed to do so, they were fined or condemned to the tower. However, in some towns, bathing was more problematic, and communities accepted that bathing would happen more frequently in the summer months, or if rainwater made it easier to provide baths.

By the 15th and 16th centuries the public bathhouses began to have poor reputations. Prostitution was a problem at baths in major cities, and during the 15th century, municipalities closed down the baths to bring an end to it. This meant that common people in towns had few ways to get clean.

THE VALUE OF PERSONAL CLEANLINESS

By the late medieval period, health manuals were written that told of the benefits of bathing, especially in soothing warm water, so people of the upper classes who could read these books valued bathing and equated cleanliness with better health. The manuals also advised hand washing before eating and recommended that people wash their hands and face when arising and to rinse out their mouths with water.

Shipping manifests indicate that soap was used during the Middle Ages. Lye would have been harsh but effective. Italy made milder olive oil-based soaps that would have been gentler. From the 12th century on, soap makers became very important in Britain. At that time soap was created from a mixture of fats and ashes. (Today's form of glycerin soap was not created until the 19th century.) As the Romans did, people of medieval times also used scrapers as an additional method for cleaning the body.

For royalty or lay nobility, dinner was not served in the great hall unless bowls were passed so that people could wash. Washing up after a meal also became customary with basins or ewers (pitchers), and servants or children brought towels to the table.

Cleanliness was made more difficult by the fact that medieval people's wardrobes were limited. Though clothing was soaked and washed occasionally, the lack of variety of outfits meant that it wasn't always possible to wash clothing and get it dry.

Personal Baths

By the 11th and 12th centuries space was provided for baths in the homes of lay nobility, but the process of bathing wasn't easy. Water for the tub generally had to be carried inside and heated over cauldrons, and unless the tub had a drainpipe of some sort, then the water needed to be carried away afterward. As a result, baths tended to be taken less frequently (generally weekly), and bathing during this time was communal so that the water needed and the work required could be put to the best use. Servants would have usually bathed outside to reduce the need for dragging in more water, and winter bathing was likely kept to a minimum because of draftiness.

Because full body baths were an ordeal, "spot" and sponge bathing of the face, hands, and feet was encouraged. Smaller basins could be used, allowing more economical use of water. While some foot bathing was ritualistic, there is also reference to how soothing it was to bathe one's feet in water after a long day walking along muddy streets or working in fields. In Ireland, offering a bath was part of hospitality. If one couldn't offer a whole bath it was proper to at least offer guests the chance to wash their hands or feet.

In hospitals and sickrooms nurses gave patients sponge baths. Hildegard of Bingen wrote about the process of face washing in her book, *Physica*: "But one whose face has hard and rough skin, made hard from the wind, should cook barley in water and, having strained that water through a cloth, should bathe his face gently with the moderately warm water. The skin will become soft and smooth, and will have a beautiful color. If a person's head has an ailment, it should be washed frequently in this water, and it will be healed." In Florence healers recommended that hands be washed in vinegar but they discouraged bathing because it opened pores to disease. (This followed the Greek theory of Democritos about the dangers of open pores.)

At monasteries, those in the infirmary were usually offered warm baths, but otherwise, attitudes toward bathing varied. Some monastic orders made bathing in hot air and steam part of a regular bathing regimen while others forbid bathing except at Christmas and Easter. Elsewhere, bathing was viewed as self-indulgent, and as a result, some monks used bath refusal as a form of self-denial. Cold baths were viewed as preferable, as they were thought to dampen fleshy desires.

For royalty, plumbing was refined for both elegance and comfort. In England in the 14th century, the king had both hot and cold faucets for his bath. Windsor Castle must have lacked running hot water, as there is record of large earthenware pots being used to heat water by the furnace and then fill the king's bath, which would have been a very large wooden tub that could be rolled into place. Some tubs had fabric liners to make them more comfortable to sit in; others had fabric canopies to shield bathers from drafts.

By the late Middle Ages the nobility had reverted to Roman ways and again bathing was a luxurious communal experience. In the late 1400s, a Bavarian king ordered preparation of five meat dishes to be enjoyed in the bath.

PUBLIC HEALTH AND SAFETY LAWS

By the Middle Ages public health and safety laws were of paramount importance. While an understanding of bacteria or the possibility of waste materials harboring disease was not approached from a scientific standpoint during medieval times, the unpleasant odors combined with a healthy suspicion that exposure to sewage was not a good idea led to the development of city sanitation rules.

Public health laws during the Middle Ages were largely left to local magistrates and nobles so there was no universal system or an overriding philosophy. In Venice a committee of three nobles established rules for burial, banned the sick from entering the city, and threw any "interlopers" in jail. Milan was one of the first towns to create an official government position to oversee health and safety, and in about 1410, the city put on staff a physician, a surgeon, a notary, a barber, two horsemen, three footmen, and two gravediggers. Other physicians signed on as advisors. The government plan for Milan was implemented elsewhere in Italy but northern Europe took additional time to come around on this.

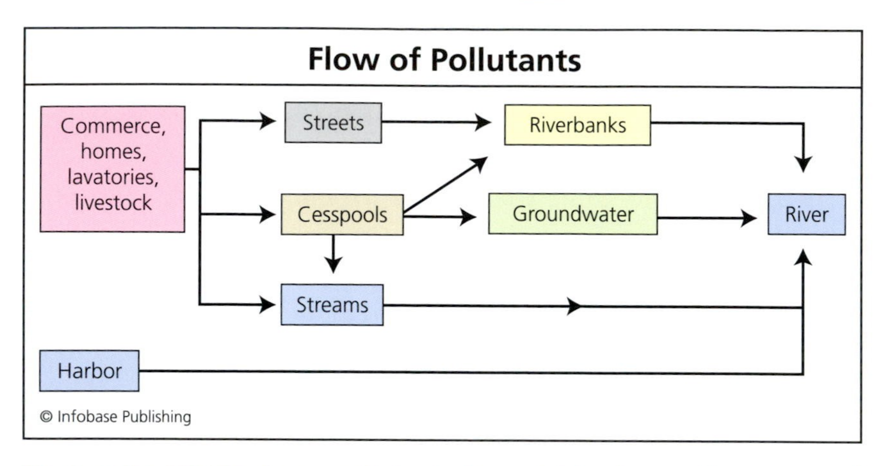

During the Middle Ages, pollutants flowed into rivers—the source of "clean" water—from many avenues that were not well understood at the time.

By the 12th century the port of Marseilles employed workers to clean the streets adjoining the harbor to keep garbage from going into the water, and the crews of all boats entering the harbor had to swear to keep the harbor clean. Garbage could not be dumped into the harbor and the hull of the ship was not to be scraped while at port. Penalties were exacted if these laws were ignored.

Lice and nit removal was also important, but that, too, bore some regulation. In the 13th century in Italy there was a civic regulation that prohibited citizens from delousing each other in public arcades.

Waste and by-products of all trades were to be disposed of properly. Zoning laws in medieval cities dictated that businesses that produced noxious smells or waste products were to be located in districts away from homes. The preferred location for these enterprises was outside the city walls and downstream from where the city water was drawn. Butchers often slaughtered the animals outside their shops, and fishmongers had to gut and scale the fish delivered to their stores. Consequently animal viscera were among the waste products generated by stores in the town. Many of the businesses worked with "treatment" products that caused noxious smells or resulted in excess waste. Tanners used caustic solutions to remove hair from skins and soften hides, and the various compounds used to tan and color the leather also smelled disgusting and would not have been desirable near public waterways.

Some cities even provided for the inspection of foods, and there were regulations regarding the sale of some foodstuffs. Some of it had to do with pricing and ensuring accurate measurement but other regulations had to do with not selling tainted food. One regulation was that fresh food was supposed to be sold before sunset so that customers would have adequate light to inspect their possible purchases. Meats were only to be sold in designated markets and they were subject to random inspection by city officials. Meat pies had to be made with fresh meat or poultry; stale meat could not be used. There were consequences of selling spoiled food.

Fines were common but other punishments are also recorded in community documents: Someone selling bad food might have his wares confiscated and burned. One baker who was caught selling loaves of bread that weighed less than he stated was punished by having a loaf tied around his neck; he was then paraded through the streets and his offense was explained to anyone seen along the way. This type of humiliation was designed to drive away customers since no one would want to do business with an unscrupulous tradesperson. While regulation of food is difficult today and must have been even more challenging during the Middle Ages, it is significant that rules of this type existed.

During the latter half of the Middle Ages, some cities had begun to employ physicians who were part of the municipal office that regulated trade. These city doctors had several responsibilities beyond their primary role of supervising the people who were practicing medicine in the city. They also conducted health and sanitation inspections, supervised quarantine for contagious illnesses, and advised the city government during times of crisis such as an outbreak of the plague. There were two occasions when their clinical skills were called in to play: If a person died of suspicious causes, these physicians sometimes functioned as coroners to more clearly identify the cause of death, and in some communities they cared for the sick who were too poor to pay for any other medical care. A public physician called a *medici condotti* was employed in the Italian city of Reggio in 1211 to assist with inquests, treat plague, and tend injuries inflicted on prisoners.

CONCLUSION

Scientific advances to improve living habits were scant during the Middle Ages, but they weren't nonexistent. Medieval monasteries, castles, and towns mastered the art of water delivery and sanitation removal, and public laws were put in place to try to safeguard the communities and their water supplies.

While personal bathing was very difficult and the popularity of public baths waxed and waned, there was a basic understanding that cleanliness was helpful to maintaining good health, and monasteries and hospitals—while not understanding bacteria and germs—still knew that occasional hand washing was beneficial.

7

Terrifying Illnesses of Medieval Times

Until the late Middle Ages, no experience with disease had been quite so terrible as the spread of what is thought to have been the bubonic plague. This plague was a virulent illness that spread easily and killed quickly. As trade routes expanded, ships traveled from China throughout the Middle East and Europe, so disease could spread through a much wider swath of the world than ever before. The medical profession had absolutely no idea what caused this illness, and therefore had poor methods for controlling or attempting to cure it. The first outbreak occurred in the Byzantine Empire in 541–542, and it was known as the Plague of Justinian (after the Byzantine ruler of the time who also contracted it). Minor recurrences of the plague occurred until the eighth century when the illness disappeared for a time. The plague, next referred to as the Black Death, recurred in 1347–51, and during a three-year period approximately 20 million Europeans—one-quarter of the population—died from it.

Most epidemiologists believe that *Yersinia pestis* caused the Black Death. (More recently, a few experts have suggested that the plague was caused by an unknown microbe that no longer exists. A few others felt it was an Ebola-like illness or possibly a disease caused by a toxic mold.) The *Y. pestis* is a bacterium that primarily infects rodents (rats and ground squirrels). Fleas picked it up from these animals and transferred it to humans via fleabites. During a pandemic (when a disease spreads widely through a large territory), an illness may infect all in its path who are susceptible and then eventually "burn out" after reducing a population to those who have some level of immunity.

Smallpox and leprosy were two other illnesses that were greatly feared. Like the plague, smallpox was capable of killing large numbers of people, and those who survived were frequently blinded by the disease. Leprosy may have been one of the first examples of an illness that was *endemic* to one area, and then spread elsewhere along trade routes. In addition to the very noticeable ulceration of the skin, sufferers' fingers frequently became deformed. The disease did not result in immediate death, but because the symptoms were so unappealing, communities forced lepers to live in isolation.

This chapter will examine the plague, how it spread, and its social and economic impact on Europe. The effect of smallpox and the attitude and treatment of leprosy will also be discussed.

THE PLAGUE OF JUSTINIAN

The Black Death was not the first widespread illness that the world had encountered. An earlier pandemic, the Plague of Justinian, had spread through the Byzantine Empire during the sixth century (541–542 C.E.). Justinian (483–565) was a very successful emperor of the Eastern Empire who was working to reunite the Roman Empire—by 540 he had conquered all of North Africa and Italy. His armies were on their way to *Gaul* (present day northern Italy, France, Belgium, western Switzerland, and parts of the Netherlands and Germany) when they were slowed by a counterattack. Justinian's army was soon weakened by the plague. The disease started in Egypt and began to spread, probably via rats (and fleas) that traveled with the grain on trade ships, causing high mortality. From the Byzantine Empire it spread as far north as Denmark and to the west it found its way to Ireland and also traveled as far south as the northern sections of Africa. Modern historians refer to this early plague as the Plague of Justinian, because the Roman emperor also contracted the illness.

The Plague by Arnold Böcklin (1898) *(Montana State University)*

During its height, the virulence of the Plague of Justinian was almost as destructive as was the Black Death later on. According to the Byzantine historian Procopius, this plague killed 10,000 people per day in Constantinople. The accuracy of a statistic from such a long time ago can never be verified, and scholars put that number at something closer to 5,000 per day; however, there is no doubt that the death toll was heavy. Approximately 40 percent of Constantinople's inhabitants and as many as 25 percent of the population of the eastern Mediterranean may have died from this first wave of the plague. Bodies were left stacked in the open because the community ran out of space and manpower to bury them.

With this particular plague, there were less devastating recurrences in the Mediterranean basin as new generations were born, or people who lacked immunity moved to the area. Then, in about 750, there was a lull in its spread for reasons that remain unknown.

The Plague of Justinian had a major impact on history. Justinian had built a powerful empire and was on the verge of reuniting the Western Empire when the plague weakened his armies. While they were able to retake Italy, they were never again strong enough to move north.

THE SPREAD OF THE BLACK DEATH (BUBONIC PLAGUE)

After the last recurrence of a weakened form of the Plague of Justinian in 750, the plague and any other *epidemics* disappeared for almost 600 years. Then in 1331, the bubonic plague got a foothold in China, where the outbreaks caused massive death rates and economic chaos. The disease spread again by trade routes and shipping caravans to cities around the Mediterranean. On one ship it was reported that out of the 332 people who were making the journey, only 45 were alive when the ship arrived in Cairo.

As the ships docked in various harbors, the death tolls in the port cities varied but sometimes were as high as one-half to one-third of the population. By 1348, the plague reached Italy, and the

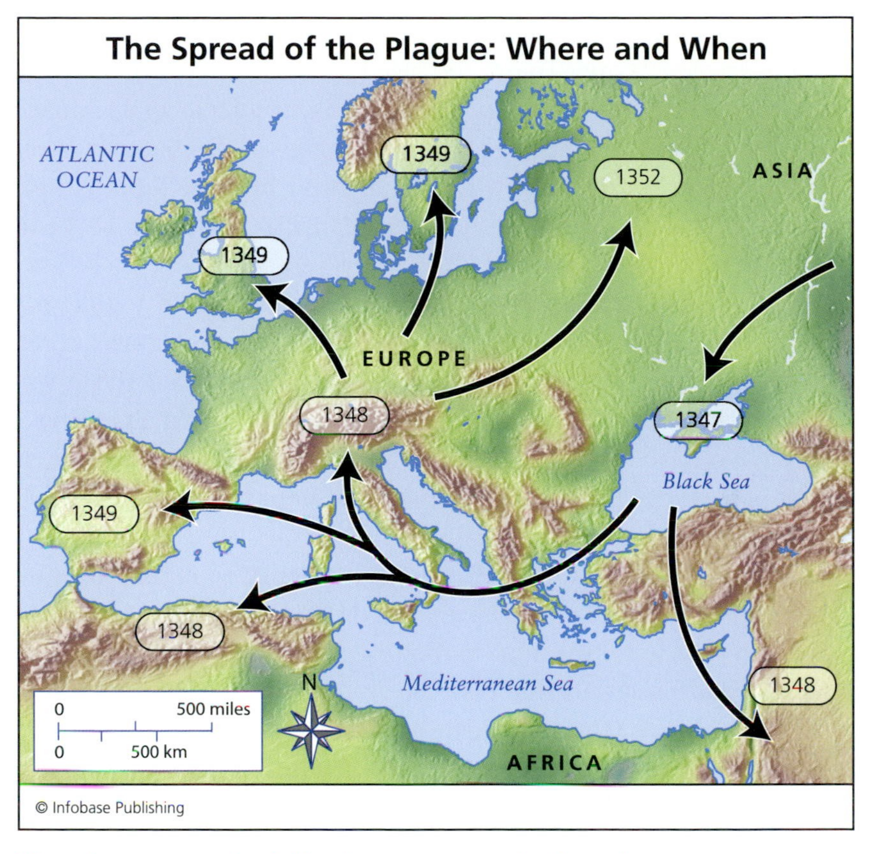

The plague spread quickly along common trade routes.

plague then started an unprecedented rampage throughout Europe. By the time the disease had petered out three years later, anywhere from 25 to 50 percent of the population had died from it.

THE NATURE OF THE PLAGUE

In China, the original form of the plague started with a very heavy nosebleed and progressed to aches, high fever, and death soon after. By the time it reached Europe the symptoms had changed, and the plague was not one simple illness; it presented itself in one of three forms. The most common variation started with swellings (buboes) in the armpit or groin; sometimes there were just a few large swellings (the size of an apple), and other times there were many small ones (more like the size of an egg). These swellings occurred all over the body, and black or purple spots caused by internal bleeding appeared on the arms or thighs and were accompanied by high fever, headache, and fatigue. Most people died within a week, and these buboes came to be known as a certain sign of death.

The second variation was of the pneumatic variety and was even more virulent. This form of the illness traveled from person to person through the air (not by fleabite), attacking the respiratory system. Those who came down with it might live only a day or two. The third version of it was *septicemic* (also known as blood poisoning). It was also lethal and people died quickly after contracting it.

Because of the strong belief held by many that religion and health were connected, some explained the disease as one of God's ways to scare mankind into banishing sins. A few attributed the plague to poisonous air that resulted from a recent earthquake, and still others looked for scapegoats among the people around them. Throughout Europe, it became relatively common to blame Jewish

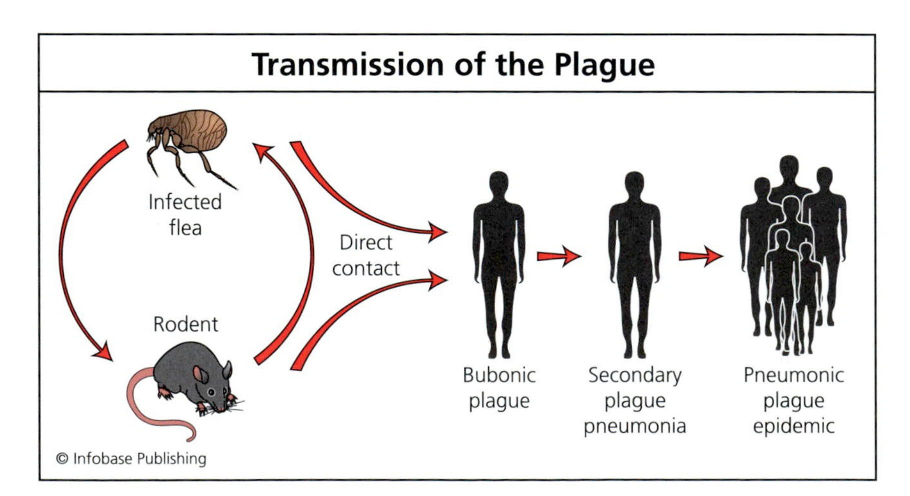

Fleas became infected by rodents carrying the plague, and the fleas transmitted the disease to humans.

people, lepers, and other minority groups for the plague. Some said the Jews poisoned the water supply and that was why everyone became sick. Punishing the scapegoats was popular in some regions, and many innocent people were burned alive. In Strasbourg (now France), 2,000 were persecuted; in Mainz (now Germany), 12,000 were killed as punishment for "causing" the disease.

One fellow, John Jacobus, a doctor of medicine who taught at Montpellier, put forth the idea that the disease likely came from dirty toilets or standing water. While this was not the way that the bubonic plague traveled best, Jacobus's reasoning was certainly more sound than that of those who believed that illness was sent as punishment.

When a second wave of the plague arrived in England in 1361, the illness primarily affected children who were not yet born when the first wave occurred. Medical professionals knew nothing of how older people could have built up immunity, so they suggested that these children had become ill because they had dishonored their parents in some way.

ATTEMPTED TREATMENTS AND EFFORTS AT PREVENTION

Rulers, noblemen, and peasants were all aware of the seriousness of the situation when the plague was identified in the community, but no one knew what to do. Massive numbers of people were affected and there were usually not enough healers, so medieval people turned to the methods that were available to them.

Many believed the illness was punishment from God, so praying was a logical option. Some communities sponsored major prayer gatherings, and priests led big processions, asking forgiveness and for God to take away the disease. Over time, community governments began to note that after a large gathering, even a religious one, sometimes even more people fell ill. Eventually Venice banned any type of gathering, and in 1523 and 1529 there were cases of towns locking up the churches so that groups of people could not congregate.

In addition to prayer gatherings, pilgrimages and offerings of gifts, often candles, were thought to be helpful. Some people made life-size candles and lit them in church as offerings to God. In Barcelona, they made a candle four miles long to surround the city and protect it. Some felt that fasting and flagellation (beating) would help. In some parts of Europe, groups of men traveled from town to town whipping one another. They believed this activity would keep everyone healthy.

Like many illnesses of the period, the plague had its own patron saint: St. Roch. He lived in the 14th century and suffered the plague but overcame it and went on to cure others. This provided inspiration for many communities.

Ibn Battuta, a 14th century Muslim traveler from Morocco, visited Damascus and wrote that the city was brought to a halt by illness, with 2,000 people per day dying at the height of the sickness. The Syrians determined that fasting and prayer were the best remedies, and Battuta reported that the people fasted for three successive days and then Jews, Christians, and Muslims all assembled together in the great mosque until the temple was filled to overflowing. They all spent the night there together, praying to God to spare them.

The medical community had other ideas about ways to treat the disease, but there were not enough healers once an epidemic hit a town. Guy de Chauliac, the most celebrated surgeon of the 14th century, recommended purging and bleeding but of course, this further weakened those who were ill.

Preventive Measures

If the plague came to a community, it was a dire situation, and superstitions thrived as people grasped at any suggestion offered in order to stay well. During the 14th century people believed that using leeches, bathing in human urine, wearing excrement, placing dead animals in homes, and drinking molten gold and powdered emeralds could safeguard them from illness. Another preventive measure involved killing local dogs and cats, thinking that they somehow were causing the illness to spread. The role of rats was not known at this time.

Some realized that lack of exposure reduced likelihood of contagion, so they tried to avoid being near those who were sick. When possible, plague patients were sent to pesthouses to live in isolation from other people. Among the healthy, many left town at the first sign of illness in the community, and they went abroad or to the country to get away. The safest place to be was in a remote village or on a farm away from the major population.

Those who remained in a community needed to believe they would be safe, so theories on lifestyle issues were put forward: Some felt that modest living was protective. Others felt that it was hopeless, so they might as well spend their final days dancing and drinking.

John I, Duke of Burgundy, wrote on avoiding the plague in 1365: "Avoid too much eating and drinking and avoid baths which open the pores, for the pores are doorways through which poisonous air can enter the body. In cold or rainy weather, light fires in your room. In foggy or windy weather, inhale perfumes every morning before leaving home. If the plague arrives during hot weather, eat cold things rather than hot and drink more than you eat. Be sparing with hot substances such as pepper, garlic, onions and everything else that generates excessive heat and instead use cucumbers, fennel and spinach."

Italy began to require that travelers to Italy first spend 30 days in present-day Dubrovnik. Those who remained healthy after this period were permitted to enter the country. Eventually this length of time was extended to 40 days (Italian word is *quarantenaria*) and from this has come the word *quarantine.*

In Milan, the town council ordered that the homes of the sick should be sealed. While this was a heartless way to treat an ailing person and his or her family, the city did see positive results. By enforcing isolation to reduce the possibility of contagion, only 15 percent of Milan's population died during the outbreak.

While some healers and physicians almost certainly left town when they got word of the plague nearby, others stayed to tend to the sick. They had two primary protective methods that they used. One involved covering their bodies fully by wearing long leather gowns. Their faces were covered with masks that featured birdlike beaks. The beaks were designed to hold aromatic herbs, which they believed protected them from catching the plague.

Another method used by healers involved fumigating the air when they visited a sick person. They carried portable incense burners filled with mixtures that were thought to counteract the "vapors" of the plague. If neither the body protection or the incense burner was available, then a healer who visited a sick person might sniff amber-scented nosegays and pomanders after the fact; these were supposed to have cleansing properties.

BREAKDOWN OF SOCIAL ORDER DURING THE BLACK DEATH

The Black Death created great fear within communities, and most towns experienced a breakdown in social order with very few people willing to help the ailing. Hospitals were primarily for travelers or the poor, so there were not many facilities that would take in the sick, thus people remained in their homes with only family members to care for them. Neighbors refused to help other neighbors, and with the death rate astoundingly high, the last surviving family member was often left with no one at all to care for him or her. In a few cases, wealthy families were able to entice servants to remain with them by offering higher wages, but of course, most of the servants eventually contracted the plague and died. As the disease continued to devastate towns, many people simply lost hope and committed suicide.

Extra burial grounds were needed, and in each location, the number of healthy men to carry out burials dwindled over time. At first people brought bodies of loved ones to the churches, but soon the numbers became so overwhelming that even that effort became too much. They eventually resorted to mass graves for disposing of as many bodies as possible.

ECONOMIC AND POLITICAL IMPACT

For 800 years prior to the coming of the Black Death, the population of Europe had grown steadily and the economy improved. However, the large population brought pressure on the agricultural needs of the area and this resulted in high food prices in the market. Landholders did well, and peasants fared poorly since they needed to eat but could not afford to break away and support themselves if they weren't working on noblemen's lands. It became easy for the landholders to oppress the poor and make their lives terrible. While the wealthy lived well, the peasants were paid poorly, they were fed poorly, and had little or no access to any form of healer if they became ill.

The plague killed so many, particularly among the serving class, that the balance of power shifted a bit. A good example of what happened is provided by England. When the Black Death came to England in ca. 1348–49, with subsequent plagues to come, the illnesses were so destructive that the population of the 1370s was between one-half and two-thirds of what it had been formerly, remaining at about 2–3 million people for the next century. As a result of this change, the peasantry and gentry (the class below the nobility) began to gain a new level of power. As the need for food and living space diminished, rural rents dropped, and this provided peasants with the opportunity to choose living space for themselves. Increased peasant wealth and a shortage of labor meant that landowners found it more difficult to find workers and could not be as oppressive in their demands.

Over time, a new pattern emerged. The custom developed that if a person worked for 10 years as a servant, he or

(continues)

(continued)

she should then be released. The poor married later since they worked as servants early on, and they had fewer children because they started married life late. This caused a decline in population. Because of these changes, the landowners had difficulty making the land profitable, so they turned to a new arrangement that involved leasing out the land and letting each tenant worry about how to make it profitable.

The economic shift began to lead to a political shift as well. When the poorer classes rose, the ruling classes began to provide for more popular participation. Eventually, the formerly landless gentry began to take on positions of municipal leadership, and they were more likely to have access to healers or barber-surgeons. This introduced a new level of popular participation in local and national government. This system paved the way for the broader level of representation in governments like Britain (and eventually the United States) today.

In 1350 the Italian writer Petrarch (Francesco Petrarca, 1304–1374), noted: ". . . [it] looks to me as if the end of the world is at hand."

SMALLPOX AND LEPROSY: TWO OTHER SCOURGES

Smallpox and leprosy were two illnesses that occurred frequently during this period and were greatly feared for different reasons. Smallpox was often deadly, and leprosy was slower to kill but it was terribly disfiguring. Communities took serious measures to halt the progression of both diseases.

Smallpox

Smallpox dates to at least 10,000 B.C.E., and, like the bubonic plague, it also spread in epidemic-style waves, wiping out big populations or leaving survivors incapacitated, often from blindness. The symptoms of smallpox appear suddenly and include a high fever, chills, headaches, back pain, nausea, and vomiting. After two to four days of fever, very noticeable red *pustules* erupted on the face and body; some could appear within the eyes, affecting vision.

The first person to write a medical description of smallpox was Razi (see chapter 8) in about 910 C.E. A well-respected doctor in the Eastern Empire, he noted that the disease seemed to be transmitted from person to person and that those who survived acquired immunity. This led to experimentation with early forms of *vaccinations.* People seemed to realize that if a person could be "given" a mild case of an illness, it would prevent them from becoming deathly ill with it later on. People in Asia learned to infect people with small amounts of smallpox to lessen the likelihood of illness. In China, powdered scabs of smallpox blisters were blown through a tube into the nostrils of healthy people. Later on, the Chinese created a pill from fleas removed from cows (cowpox was thought to be related to smallpox) to prevent smallpox. In India, the scabs or pus from a person with smallpox was scratched into the skin of a healthy person. These techniques were taught to other cultures by the caravan travelers, who passed on the knowledge they gained as they traveled. The system was used in the Eastern Empire and eventually introduced in Europe. The actual invention of the vaccination process is attributed to Edward Jenner in 1798 making it likely there was not continuous use of this process after the Middle Ages.

Leprosy

Leprosy also predated the Middle Ages and is described as early as 2400 B.C.E. by the Egyptians. It appears again in the Ebers Papyrus (1600 B.C.E.). In India it is recorded in 600 B.C.E.; it is noted as involving loss of sensation, loss of fingers, deformity, ulceration of the skin, and the sinking in of the nose. The disease seemed to have remained "local," until trade routes improved. The Greeks may have brought it from India, and then the Romans may have transported it into Europe. The disease increased heavily after the sixth century and peaked during the Middle Ages. While the disease is actually only slightly contagious, the overwhelming influx of so many people during the Crusades meant that leprosy was carried to multiple communities during this time. Because so many people returned from the Crusades with some form of illness, this was actually quite embarrassing to the church. By the 13th century, one out of every 200 Europeans suffered from leprosy making it quite a problem.

Leprosy was a terrifying illness; it killed slowly but it was so disfiguring that it was like being sentenced to a painful lingering death. The disease caused bone degeneration, finger and toe mutilation, and most notably, patches of scaly skin. Because it was so disfiguring, the disease became highly stigmatized. People of the day had many beliefs as to how it traveled. Some thought it was contagious if one stood upwind of a leper; others thought that if a leper walked barefoot through grass, the grass could then transmit leprosy to anyone who followed; still others thought it was spread via sexual contact. Laws and social customs developed to assure that those who were not already infected would not come in contact with those who had leprosy. While exact customs varied by location, the goal of social ostracization was accomplished one way or the other. In most communities, anyone with suspicious spots was to be brought to a priest to be examined. If the

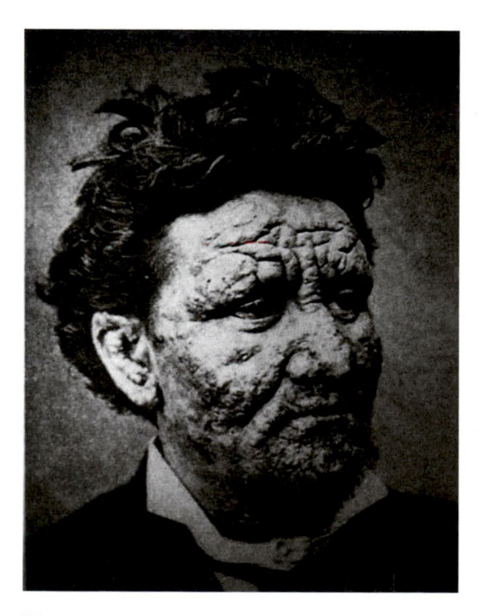

Norwegian man, 24 years old, suffering from leprosy in 1886

priest was unsure what the spots were, he could require the suspect to remain in isolation for two weeks to see if a more accurate diagnosis could be made. If a final judgment was made that the disease was leprosy, then a variety of customs came into play. In some societies, a "leper mass" was conducted at a cemetery with the leper standing in a grave, and a priest declared that the person was to be considered dead to the living. The understanding was that the person was to have no further dealings with friends or family and was only to be in contact with other lepers.

Leper laws were very strict in forbidding social contact. Lepers were required to dress distinctly so that it was clear that they were lepers; they were not to go into public places; and they were to ring a bell as they walked along or call out "unclean, unclean" so that others would have fair warning that a leper was approaching. Unmarried lepers were not to marry, they were not to touch anyone, and were to eat and drink only in the company of other lepers. Healthy people pitied lepers and would throw food to them from a safe distance.

People with other types of skin disorders (from *psoriasis* to skin cancer) were sometimes identified as having leprosy, and unfortunately for them, once they were categorized as lepers there was no escape. Even if their skin problem eventually proved to be only a rash, there was no way out once cast into a leper colony.

While a community's first priority was to protect its inhabitants, a few efforts at cures were made. Some healers believed that cooking a black snake in an earthenware pot with peppers, salt, vinegar, oil, and water made a suitable medicine for curing leprosy. It seems that those who drank this mixture became dizzy and had to be given something else to counter this side effect.

A leech book (term for a medical book from that time) from the 15th century suggests cooking a bushel of barley and a half bushel of toads in a lead cauldron and letting it simmer long enough that the meat "falls off the toad bones." This brew was then to be dried in the sun and fed to newly hatched chicks. The chicks then were to be roasted and boiled and fed to the leper in order to bring about a cure.

By the 11th century, the number of hospitals built to house lepers began to grow. By 1225, there were 19,000 leprosaria in Europe. While many of them were quite small, this large number of facilities puts in perspective the magnitude of the disease. Only 125 years later, the leprosy rates began to decline. The cause of the decline is a much disputed issue. The widespread occurrence of the Black Death may have killed so many people that it slowed leprosy. Others feel that the rising rate of tuberculosis led to a decline in leprosy as tuberculosis overpowered the leprosy germ.

The cause of leprosy was not discovered until the late 19th century, and today leprosy is called Hansen's disease after Gerhard Hansen (1841–1912) who discovered that the illness was caused by the bacillus *Mycobacterium leprae.* The new name is used in an effort by scientists to remove some of the stigma. Today nearly 15 million people still suffer from leprosy but the illness is not as contagious as was once thought so it has not gained the amount of attention that many other illnesses, such as malaria, have.

CONCLUSION

Since the medieval people had no understanding of what caused illnesses, and very little understanding of how diseases were passed from one person to another, they were helpless in the face of highly contagious illnesses such as the bubonic plague and smallpox. While some communities began to understand the benefits of not being near those who were sick, they still lacked knowledge as to how to make someone well. If a person did improve, the return to health came about because of good luck that was then explained as the answer to a prayer.

Leprosy was so disfiguring and painful that people feared it and assumed that it was highly contagious. Because people could live for years with the illness, communities began to establish customs that provided for the isolation of anyone suffering from leprosy. These options ranged from leper houses (hospitals devoted to lepers) to community-wide practices that involved socially ostracizing anyone with the disease.

8

The Golden Age of Islamic Medicine

During the Middle Ages, the two halves of the Roman Empire developed very differently. The western section moved forward in fits and starts as a feudal government system evolved, and honored a strong allegiance to the ruling elite of the Catholic Church. In contrast, the Eastern Empire grew and prospered in what has become known as the Golden Age of Islam. While this area still faced its challenges and unrest, in general, a culture of scholarship and learning was permitted to grow.

The Byzantine Empire, as it is now known, was strongly influenced by the Islamic religion founded by the prophet Muhammad (570–632) in the early seventh century. Muhammad was an orphan raised by relatives who intended him to be a merchant. At the age of 40, he received a series of visions in which he believed the Quran (Koran) was revealed to him. He determined that these visions meant that he was destined to be a prophet, and in 613 he began to preach publicly about Allah's power and goodness. In less than 25 years, almost all of Arabia accepted Islamic beliefs and became Muslims. Over the next 100 years, Islamic beliefs spread to Persia, Egypt, North Africa, and Spain. During this early period,

Jews and Christians were accepted within the Muslim community but this attitude changed when the Crusades began.

In both the Eastern and Western Empires, the spirit of the time was to honor the past, and no particular value was placed on "newness." The Islamic scholars who studied science and medicine relied on the writings of those who had preceded them in the Roman Empire. They assumed the responsibility of preserving, analyzing, and developing theories, and they did not question them. And just as the West needed to exist peacefully with the Catholic Church, the Byzantine Empire needed to blend its belief system into one that worked with the Islamic religion. These factors (honoring the past and adhering to what was acceptable to the religion) set the stage for the medical and scientific investigations of the period.

Two particular areas that benefited from developments in the East were the field of pharmacology and the introduction of the hospital as part of community development. Hospitals were originally "group housing" for the poor, lepers, or the insane, and this changed the way these societal groups were treated.

This chapter introduces Islamic medicine (also known as Yunani medicine) that developed over a 900-year period and still has followers today. This chapter describes how the Byzantine people became interested in preserving the medicine of the Greeks and how the Arabs contributed to the creation of hospitals, the cataloging of medicines, and brought about a better understanding of the workings of the heart. Several of the physicians who were particularly crucial to the development of Islamic medicine are introduced.

MEDICINE SPREADS EAST

Just as in the West, some people within the Islamic religion believed that disease was sent by God as punishment and that a certain acceptance of one's fate without medical intervention was necessary. Death and suffering were sometimes seen as part of the religious good (martyrdom). Eastern healers of this time believed illness was caused by evil spirits (jinni) and the evil eye (*al-ayn*), and that if the spirits were contacted properly, the healer could strike a bargain that would bring the patient back to good health. If an illness was to be treated, folk remedies supplemented by magic were the methods that prevailed during the early Middle Ages.

The environment for science and medicine began to change during the seventh century because Muhammad, who was soon viewed as a prophet, was to have such a profound influence on most aspects of Islamic life. He believed in education and taught that physical and spiritual well-being went hand in hand. He believed there were two areas of knowledge that had true value: knowledge of faith and knowledge of the body, and faith in God was an important part of Muhammad's philosophy. He was quoted as saying: "God sends down no malady without also sending down a cure."

Much of the medicine attributed to Muhammad had to do with sensible eating and drinking in order to maintain good health, but Muhammad paid attention to his visions and related what he learned from them to healers. Muhammad's advice ranged from the general to the very specific; he supported the use of honey as a health-restoring food and felt that cupping was good for one's health. He was known to actually use cauterization on the injured, though later he came to forbid it. He did not support the wearing of amulets if it was to bring forth supernatural powers, but he believed that amulets could be used in other ways to help ward off illnesses.

Though *folk medicine* continued to be important throughout Muhammad's lifetime, Muhammad's emphasis on education created an Islamic mind-set that was open to learning new things or relearning some old ones. Muhammad was succeeded politically by four *caliphs* (spiritual leaders of Islam) who governed until the end of the seventh century, and the occurrences that took place during this time also helped introduce the practice of Greek science, philosophy, and medicine. As these rulers conquered Syria, Persia, and Egypt, where the practice of Greek medicine was strong, the conquerors were influenced by those they conquered.

India also provided a channel through which Greek thought was transferred into the Islamic world. India had used Greek thinking as the basis for advancement in math and astronomy, and during the period of the Roman Empire, people traveled back and forth to India via land and water routes and brought with them Indian medical practices, many of which were based on Greek understanding.

Later, the city of Jundi-Shapur (in the area then known as Persia) became a center for Muslim scholars who applied themselves directly to the original Greek sources and focused on assembling the Greek texts—including Galen and Hippocrates—translating their works, and correcting and verifying earlier knowledge. The years between 800–833 became known as the "age of translations;" scholarship was held in high esteem and a great deal of progress was made. This heavy emphasis on translating the works of Greek scientists and physicians continued for the next two centuries, leading to a revival of interest in these fields.

Hunayn ibn Ishaq (809–873), one of the most important translators of this time, translated into Arabic many of Galen's medical and philosophical writings and writings of Hippocrates and Dioscorides. He was particularly interested in Galen's work, and with his students he translated 129 works into Arabic, providing more copies in Arabic than survive in Greek. This decision on what to make accessible had a major effect on Islamic medicine. Because so many of Galen's works were translated, he became a father figure for Arabic medicine.

In the mid-eighth century, Baghdad became the capital of the Islamic Empire, and both Baghdad and Cairo became important centers of scholarship; the library in Cairo (established in 988) was said to have more than 100,000 volumes. (These were destroyed after just under 300 years when the Mongols conquered Baghdad.)

As independent scholars and physicians became more interested in practicing the medicine taught by the Greeks, they faced a dilemma similar to the one faced in the West. They needed to reconcile these practices with the beliefs of those who felt that religion and prayer should outweigh all else. Those who argued for the Greek style of medicine noted that the practice of medicine was a form of service to man, and they persuaded others that the art and practice of any type of healing was second only to faith in earning God's blessing.

AL-RAZI (CA. 854–BETWEEN 923 AND 935): LEADING PHYSICIAN

Al-Razi (Rhazes; also Abu Bakr Muhammad ibn Zakariyya al-Razi) became known as the greatest physician of the Islamic world. He worked from the basic belief that when it came to looking for a cure, good sense and experience offered the ultimate authority. Al-Razi was director of the first great hospital in Baghdad, and he created a remarkable body of work, writing more than 200 medical and philosophical treatises including *Continens,* known as the *Comprehensive Book of Medicine.* Jewish physician Faraj ibn Salim (known as Farragut) translated this book into Latin circa 1279 and it was used as a reference tool in western medicine for several centuries. No Arabic copy of the work survives.

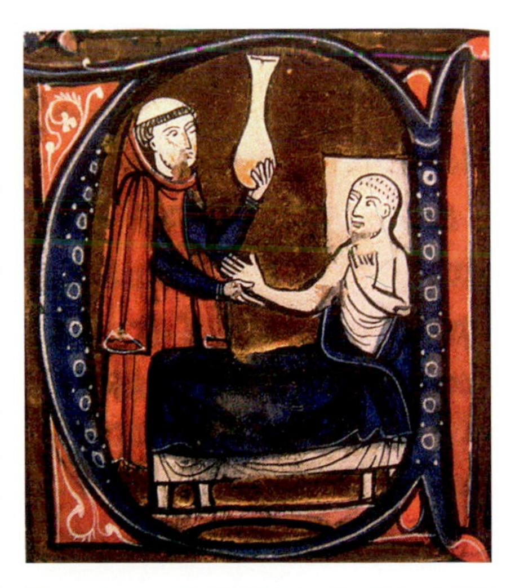

European depiction of the Arab doctor al-Razi (Rhazes), in *Recueil des traités de médicine,"* 1250–1260 by Gerardus Cremonensis

One of al-Razi's major contributions to medicine was documenting contemporary case histories. As a result of his work, scholars today can read about the signs and symptoms physicians of the time noted, the types of treatments used, and the relationship between physician and patient. In al-Razi's opinion, the physician had a responsibility to the patient for treating him or her ethically, but the patient also had an obligation: cooperation and obedience to the physician consulted. "With a learned physician and an obedient patient, sickness soon disappears," he wrote.

One story about al-Razi shows great bravery and creativity: One of his wealthy patients was suffering from a crippling ailment and was having trouble getting around. Al-Razi agreed to treat him, but stipulated that the patient's best horse and mule needed to be made available to al-Razi. Al-Razi then began a treatment that involved adding various substances to the water as the patient sat in a hot bath. After al-Razi administered some of the remedies, he suddenly brandished a knife and started shouting insults at the patient, and then he ran outside and departed on one of the fellow's steeds. The patient was so angry that he scrambled out of the tub and chased after the physician. Al-Razi later returned and explained that his actions were all part of the cure to get the fellow moving again. The fellow calmed down, expressed admiration for the technique, and rewarded al-Razi with gifts.

Al-Razi also offered one of the earliest written reports on an allergy and what he recommended for a cure. He writes of a case where someone reported getting a runny nose and watery eyes each spring and recommends that the patient avoid "aromatics" such as flowers and onions, and if that did not work, then al-Razi suggested bloodletting from the neck.

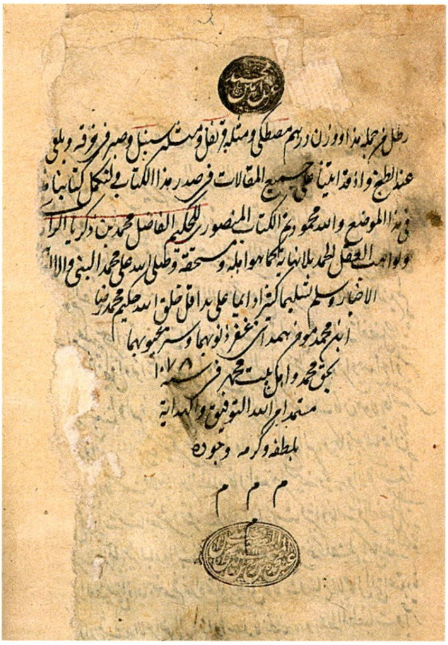

Al-Razi's book of medicine

Al-Razi wrote a separate work about measles and smallpox. He recognized them as two separate illnesses but cautioned physicians to hold off a bit on making a diagnosis because if a judgment was made too early, there was a high likelihood of making an erroneous call. Measles were caused by bilious blood and recognized as being less serious than smallpox, which could cause blindness. In smallpox, pustules that became hard and warty were supposed to indicate that the patient would die. Al-Razi wrote that proper management could help lessen the seriousness of the disease, and he recommended encouraging eruption of the pox by wrapping, rubbing, steaming, purging, and bleeding the patient. He also offered advice on removing pockmarks from the skin, since those who survived were often quite disfigured from the scarring. These remedies involved using sheep's dung, vinegar, sesame oil, and the liquid found in the hoof of a roasted ram. While they may not have been very effective, the treatments must have given both physician and patient a feeling of control.

Al-Razi had many talents. He was also a poet, a mathematician, and a chemist. It was his efforts at alchemy that ultimately brought him down. Al-Razi had promised the reigning ruler that he would turn base metal into gold. When he failed, the ruler reportedly struck him.

IBN SINA (980–1037): A PROLIFIC EDUCATOR

Ibn Sina (also Avicenna; Abu Ali al-Husayn ibn Abdallah ibn Sina) was born about 50 years after al-Razi and is sometimes referred to as the "Prince of Physicians." Ibn Sina was the son of a Persian tax collector, and he wrote in his autobiography that he amazed his father and tutors by mastering the Quran by the age of 10. He went on to study philosophy, the natural sciences, and medicine, and he gained additional education and experience and worked as a jurist, a teacher of philosophy, an administrator, and a physician. He was also the first scholar to create a complete philosophical system in Arabic. Ibn Sina soon realized that medicine was best studied via practical experience, so he started seeing patients, preferably wealthy ones who supported his work and his teachings.

The Islamic Empire and the Area Where Ibn Sina Lived and Worked

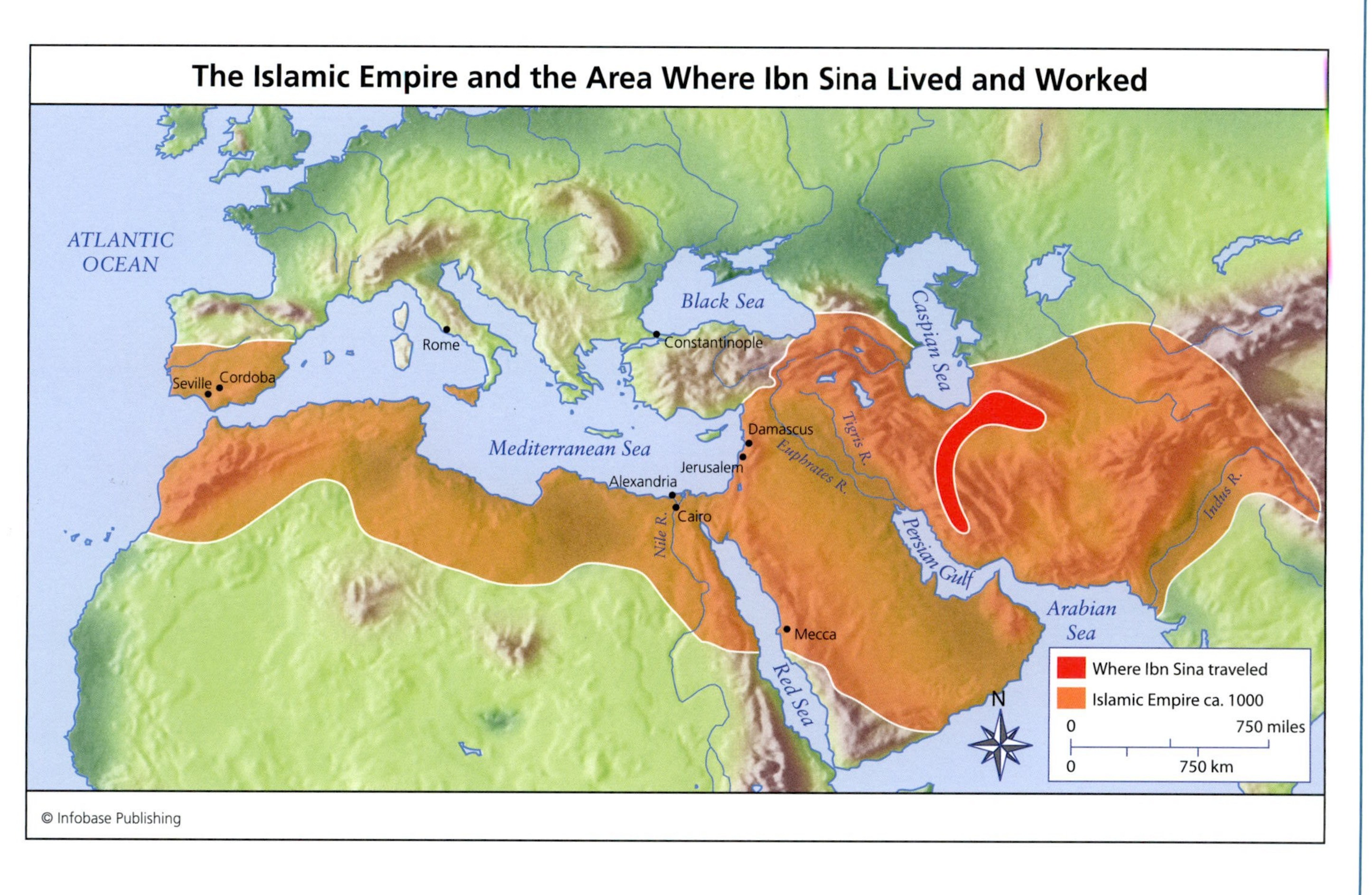

He wrote 270 works, and two of them—one on science and one on medicine—were quite extensive. As scientists have gone back to study Ibn Sina, they have found that he was a man who made important discoveries that had to be "rediscovered" centuries later. Among the amazing findings that have been uncovered are a description of Newton's First Law of Motion, a full 600 years before Newton, and also a treatise on time and motion, where Ibn Sina reached the same conclusion as Einstein did in 1905. He also wrote of the theory of the elements.

The *Canon of Medicine* was a major treatise Ibn Sina wrote for physicians. The work reflected an appreciation of Galen with Aristotelian logic. Healers used the *Canon* as a reference tool both for diagnosis and treatment, and it was used as the basis for medical courses from 1250–1600. (A greatly abridged version of the *Canon* was called the *Poem on Medicine,* and it was created to provide laymen with access to medical theory.) Ibn Sina spelled out what to look for when inspecting urine and how to evaluate the strength, elasticity, rhythm, and speed of the pulse. He wrote about the anatomy, described disease symptoms and noted practical treatments. He also provided a detailed listing of some 760 drugs as well as offering lifestyle guidance starting with appropriate care for infants right through to recommending the best regimen to keep the elderly healthy. Soothing olive oil baths were one remedy that was offered as helpful with joint pain but if the pain was particularly bad because of damp weather, then boiling a lizard in the oil would make the bath more effective.

Ibn Sina also realized that a good physician needed to be able to assess water quality. Bad water could cause disease but water that contained iron was actually helpful with stopping diarrhea and strengthening the internal organs. Ibn Sina wrote of the relationship between mind and body, and his treatments were frequently based on how one could affect the other. Surgery was also covered

(Opposite) Though Ibn Sina traveled a great deal, the territory he covered was only a small portion of the Islamic Empire of that time.

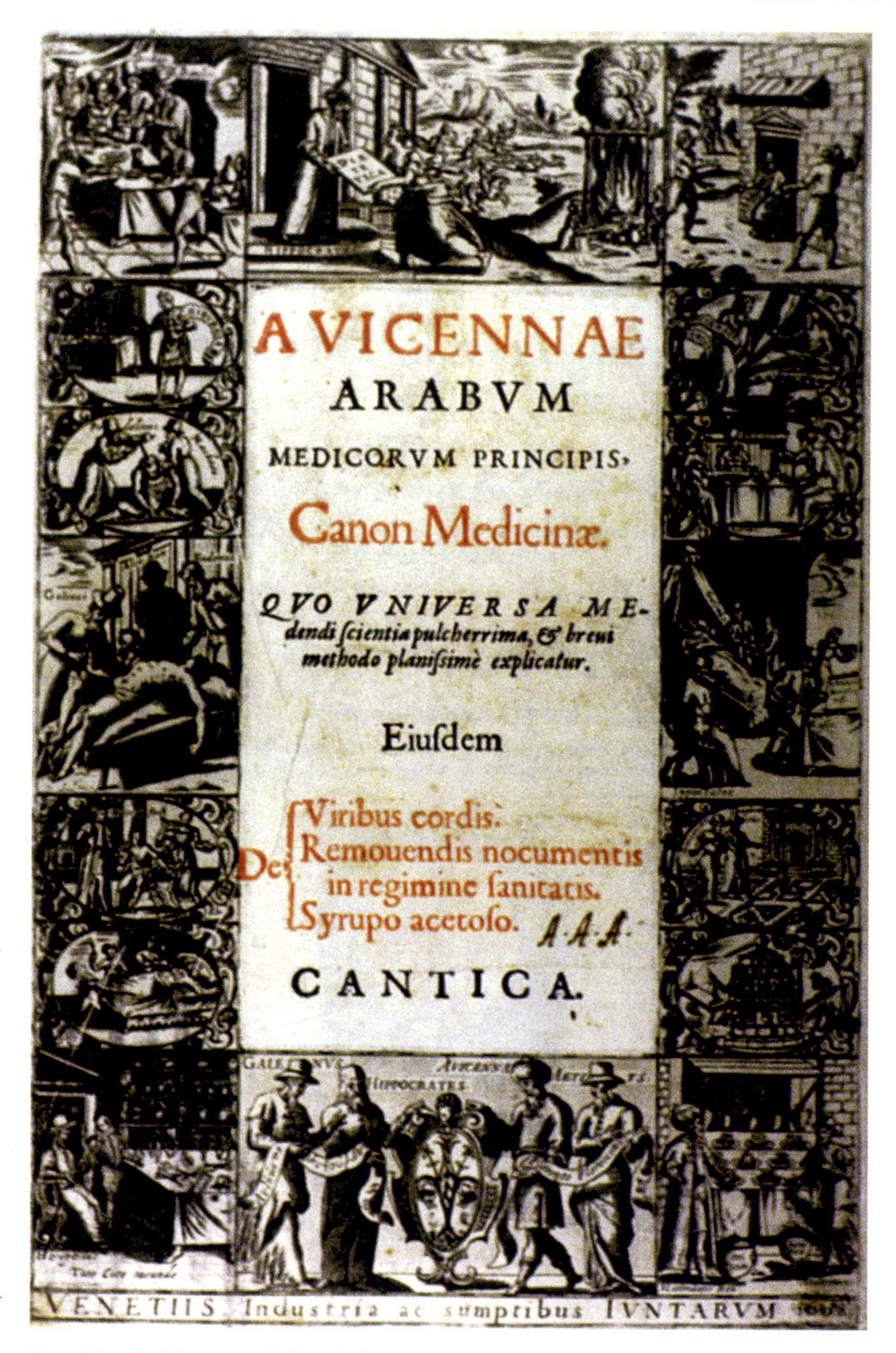

AVICENNAE

ARABVM

MEDICORVM PRINCIPIS,

Canon Medicinæ.

QVO VNIVERSA ME-
dendi ſcientia pulcherrima, & breui
methodo planiſſimè explicatur.

Eiuſdem

De { Viribus cordis.
Remouendis nocumentis
in regimine ſanitatis.
Syrupo acetoſo.

CANTICA.

VENETIIS Industria ac sumptibus IVNTARVM

Ibn Sina's *Canon of Medicine* was a very important work written for healers of the day.

in the *Canon,* including an extensive section on bloodletting, but Ibn Sina did not cover most areas of surgery in the depth that Abu al-Qasim al-Zahrawi did. (See the sidebar "Abu al-Qasim al-Zahrawi [ca. 936–1013]: Contributed to Surgical Knowledge" on page 128.)

In addition to being devoted to his work, Ibn Sina was said to have had a great love of wine and women, and when the alcohol began to affect his health, he began to treat himself with multiple medicated enemas that he took throughout the day. The process caused ulcers, seizures, and extreme weakness, and Ibn Sina is thought to have died from his own "cure."

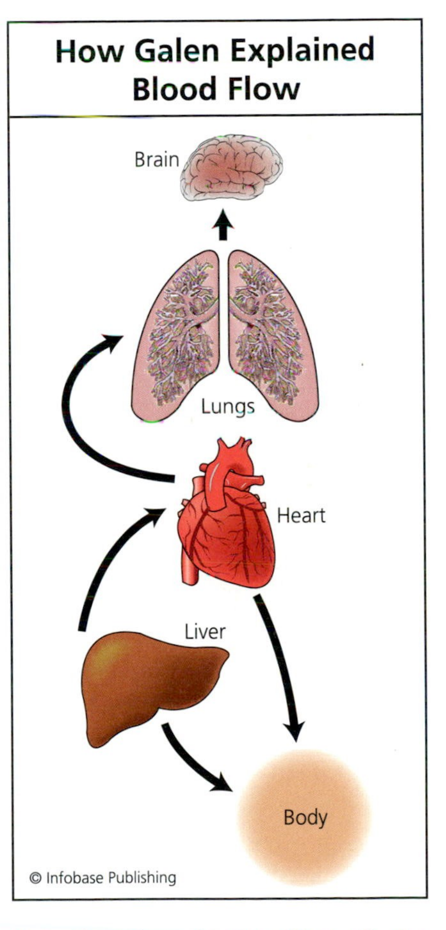

Galen's misunderstanding of the flow of blood was still the prevailing theory of the time.

WORKINGS OF THE HEART

Islamic medicine has long been valued for preserving the beliefs of the Greeks and the Romans and to some extent perfecting them, but less has been said about any Islamic medical advances. This was primarily because the most significant advance was made by physician Ibn an-Nafis and had to be rediscovered in the 20th century. In 1924, an Egyptian physician, Dr. Muhyi ad-Din at-Tatawi, wrote his doctoral thesis on some little-known writings of Ibn an-Nafis (1210–80). This thesis later came to the attention of historian Max Meyerhof, who undertook the study of what Ibn an-Nafis believed.

Though no contemporary writers of his time seemed to follow Ibn an-Nafis's lead (though some may have who have not yet been translated), Ibn an-Nafis differed with Galen on some aspects of

ABU AL-QASIM AL-ZAHRAWI (CA. 936–1013): *Contributed to Surgical Knowledge*

Abu al-Qasim (Albucasis) was one of the most respected of Islamic physicians, and his advice was listened to by physicians throughout the empire. Like other physicians of the day, he warned against taking on cases where the patient appeared to be terminally ill; he recommended that the fate of those cases be left to divine providence. Abu al-Qasim wrote one of the first comprehensive illustrated books on surgery, *On Surgery and Surgical Instruments,* and was the only Arab who seemed fully experienced with surgery.

In his book, he wrote extensively about bleeding, cupping, and cauterization. Despite Muhammad's position against cautery, Abu al-Qasim advocated its use for almost every condition, from hemorrhaging to headaches, and he provided very specific instructions. For example, for any sort of head pain, the patient's head was shaved and the cautery tool was applied to the bridge of the nose right between the eyes. The cauterization was considered complete when the bone was exposed. If headaches continued then bleeding from the arteries was recommended.

Both Abu al-Qasim and Ibn Sina agreed with Galen and supported the use of bloodletting. Even in cases of hemorrhaging, it was believed that further bleeding was helpful because it diverted the blood to the opposite side of the body. The physicians' specified 30 sites for venesection, 16 of which were in the head, five in the arms and hands, and three in the legs and feet. To evaluate the quantity of blood the workings of the heart. While Ibn an-Nafis agreed with Galen's theory that the left ventricle contained vital spirit while the right ventricle contained blood, he contradicted Galen's theory about

to be taken, the color of the blood and the patient's strength was to be considered. Generally bleeding was carried out in several small installments but if a person had "hot blood" and a fever, then it was recommended that he be bled until he fainted. Abu al-Qasim wrote that the physician should keep his finger on the pulse during the process to monitor when the patient fainted so the process could be halted before the patient actually died. His book also contained a major section on leeches; leeches were considered particularly good for bleeding from deep tissue areas that were difficult to reach with a cupping bleeding procedure. However, the selection of the appropriate leeches was an art. The wrong type of leech could cause *inflammation* and even paralysis, so Abu al-Qasim's book described the way to select the best leech for the task.

In his book Abu al-Qasim included illustrations of 200 surgical instruments, many of which may have been created by him. Some of what he accomplished is being rediscovered now by scholars who have returned to re-examine the original documents. When copyists of the time transcribed Abu al-Qasim's information into Latin, they sometimes made mistakes, and when they copied the illustrations, they often did not understand the purpose of the surgical instrument. They did not know what was important, and sometimes added decorations or detail, that, of course, changed the interpretation of the surgical tool.

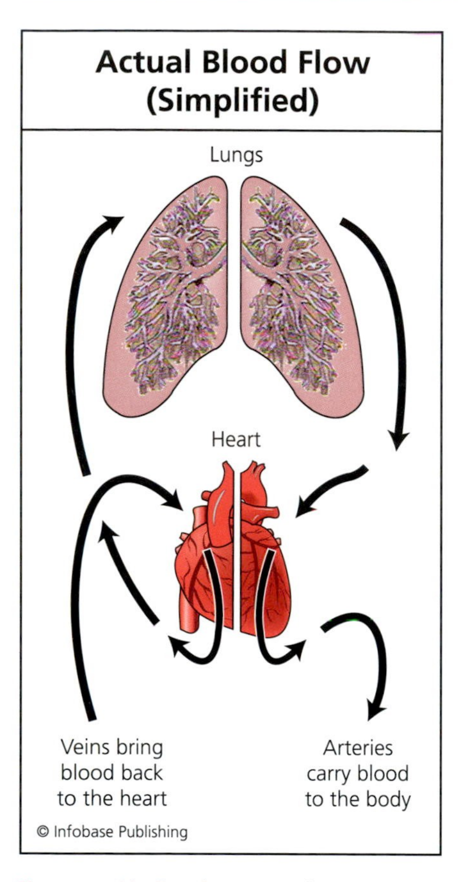

Ibn an-Nafis (1210–80) came to understand that Galen's theory of the workings of the heart was inaccurate, but his advances had to be rediscovered in the 20th century as his influence did not live on beyond his time.

how pores within the septum permitted blood and spirit to pass between the two areas. Ibn an-Nafis came to understand that the blood needed to travel from the right ventricle to the lungs to acquire air, and then return to the left ventricle. This theory—not Galen's—was correct, and it preceded Michael Servetus and Realdo Colombo who are credited with discovering this.

HOSPITALS, PHARMACIES, AND ADVANCES IN MEDICAL TRAINING

During his lifetime, Muhammad taught great compassion and he made it a point to always visit the sick to provide comfort and hope. From this, an Islamic tradition of charity grew, and charitable institutions established hospitals and educational centers as early as the eighth century, and eventually, hospitals became one of the great achievements of medieval Islamic society. These facilities tended to be large and were used for several purposes. Sometimes they served as a place for the poor, sometimes as a retirement home for the elderly, and sometimes as a place for convalescence. While they were not necessarily religious institutions, they generally had a mosque or some sort of area for prayer. Many were open to all, male and female, civilian and military, adult or child, rich or poor, Muslim and non-Muslim.

The earliest documentation that reflects the building of a hospital dates to the ninth century in Baghdad. During the next 100 years five more hospitals were built in that city. By the 11th century, there were large hospitals in every major Muslim town, and instruction in medicine was part of the hospital culture.

Pantokrator Hospital in Constantinople was founded in 1136 by Byzantine emperor John II as a part of a religious complex that included a monastery and an old folks home as well as a leper house (located four miles away from the main facility). The hospital provided a high level of patient care, and this was one of the facilities that was observed by those who came on the Crusades and contributed to what was taken back to other parts of Europe.

The development of hospitals was to change the care of the poor, the mentally disadvantaged, and eventually the sick. The Islamic people began to understand the importance of this care for a community, and over time, they created mobile units that could be taken to serve people in rural communities and organized physicians and female healers to visit the sick in prisons. However, most wealthy individuals would not have gone to a hospital but would have received private care at home.

The Development of Pharmacies and Pharmacological Knowledge

The very first *pharmacy* to exist was built in what is now Dubrovnik, Croatia. It first opened in 1317 and is still a working pharmacy today. The understanding of pharmaceutical knowledge as a separate area from other types of healing was to take Islamic practitioners far in making great progress in pharmacology.

Islamic pharmacological practitioners discovered and catalogued thousands of new drugs, greatly improving the systematic methods for processing and using medications. (The word *drug* is actually from the Arabic, as are "alcohol," "alkali," and "syrup.") Working from the Greek treatise on medicine written by

Image of a 14th-century school by Laurentius de Voltolina *(The Yorck Project)*

Dioscorides in the first century, scholars translated Dioscorides' advice into Arabic and later added to it, and eventually this work was translated into Latin so that even more people could benefit from what had been learned.

In creating new medicines, the Islamic pharmacists used ingredients from as far away as China, southeast Asia, the Himalayas, southern India, and Africa, and they introduced new medicines of their own including benzoin, camphor, myrrh, musk laudanum (an alcohol solution of opium), naphtha, senna, and alcohol itself.

The Arabs also created a new form of jar for holding medicines. It was known as an albarello and was used for storing herbs, roots, seeds, spices, and other medicinal substances. The jar had slightly concave sides, which made it easy to remove from the shelves. This was helpful in a time when medicines were created to specification and were concoctions made of many ingredients. By the 15th century, the design became popular in Europe as well.

Physicians and their Training

Islamic physicians usually obtained their training by studying under a specific doctor, and certificates of accomplishment were issued directly by the instructing physician or scholar, but it was also common for a young person to obtain various types of training, traveling to several different cities to study different subjects under various wise men. The education of doctors in the Byzantine Empire sometimes took place in a hospital setting even though each individual doctor conducted the training himself. Physicians came to appreciate the benefits of having a hospital library, and they also began using patient treatments for teaching, a concept that was not used elsewhere for a long time.

Women were permitted to train as nurses, midwives, and gynecologists, and their education was conducted in a private one-on-one setting with a tutor. Like women in Western society during the Middle Ages, women were not permitted to be treated by male doctors, so women healers filled an important role.

The death of a patient in a Baghdad hospital in 931 was a turning point in Islamic medical care, giving rise to a new interest in establishing methods to supervise and test doctors. (Formal testing of pharmacists began almost a century before they began testing doctors.) While there was common interest in tightening the provisions for medical practitioners, there was little standardization of the process. Handbooks were written for laypeople that were much like self-help books of today and instructed patients in methods they could use to judge whether the physician they were seeing was a well-trained individual or a fake. For example, one example of the type of "test" given to determine the expertise of a physician described a fellow who submitted mule urine to a physician to see if he would realize that it was from an animal not a human.

The Islamic doctor traditionally wore a white shirt and cloak, a distinctive turban, and carried a silver-headed stick. Generally the doctor was perfumed with rose water, camphor, and sandalwood. Islamic physicians were generally successful financially. They believed in charging high fees to the wealthy so that they could afford to treat the poor at little or no cost. A few used their money to build hospitals or clinics but some just lived well.

THE SPREAD OF ARABIC INFLUENCE

By the 12th century, European physicians understood that the Arabic practitioners were outdistancing them, and they began to seek out Arabic texts and translate them into Latin. This development was to have great influence on what was happening in the West. Europe's slow recovery from the fall of the Roman Empire meant that physicians had little time to preserve earlier practices because most areas were dealing with a great deal of chaos. Since the eastern countries gained control more quickly, the Arabs had the opportunity to retrieve and translate Greek texts that had been preserved, and so as the West turned to the East, they returned to medical theories that were still based on humoral balance; the standard treatments for illness were still cautery and bleeding.

As indicated in chapter 2, one of the notable translators from Arabic into Latin was Constantinus Africanus (1020–87) who smuggled out a copy of Ibn Sina's works and established himself in Salerno where he was able to share his growing body of knowledge. In addition to the translating that took place to spread information about Greek and Islamic practices, travelers also talked of the treatments they observed and noted the types of instruments used by Islamic practitioners.

CONCLUSION

By the end of the 15th century, the Islamic world had become very fragmented, which led to a drop in medical and scientific progress. The hospitals that had grown up during the earlier era became dependent upon charitable endowments for their maintenance and with time, as these funds became insufficient, the hospitals deteriorated and fell into disuse.

But Islamic medicine (now sometimes referred to as *Yunani* medicine) never fully died out and actually continued to develop long after the Middle Ages. Western medicine has been added to the medical traditions of this part of the world, but some of the practices of Yunani medicine continue. There are still schools that specialize in Yunani teachings, though more and more areas are becoming Westernized.

286 C.E.	Eastern Empire splits from the Roman Empire
ca. 400–500	Collapse of the Roman Empire in western Europe
ca. 540–604	Pope Gregory stresses that prayer is more important than medicine
541–542	The Plague of Justinian spreads through the Byzantine Empire
570–632	Life span of Muhammad, who founded the Islamic religion
Eighth century	Baghdad becomes the capital of the Islamic Empire
805	Founding of the first hospital in Baghdad
809–875	Life span of Hunayn ibn Ishaq, who translates many of Galen's medical and philosophical writings into Arabic
ca. 836	Abu al-Qasim al-Zahrawi writes the first comprehensive illustrated books on surgery
865–935	Al-Razi is known as the greatest physician of the Islamic world. He serves as director of the first great hospital in Baghdad; he writes more than 200 medical and philosophical treatises, including *Comprehensive Book of Medicine,* known as the *Continens.*
980–1037	Ibn Sina, referred to as "Prince of Physicians" writes 270 works including the major treatise the *Canon of Medicine,* which is used as the basis for medical courses from 1250 to 1600
1010–1087	Constantinus Africanus translates works of Greek and Roman philosophers and physicians into Latin
ca. 1095–1300	The Crusades

12th century	Women are permitted to teach and practice medicine at Salerno
	Many hospitals (offering food and shelter) are established for people traveling on the Crusades
	Some monasteries and castles create methods for flushing without the tides by constructing large elevated cisterns. When the water was released there was enough pressure to carry away waste.
1123	Founding of St. Bartholomew and St. Thomas hospitals in London
1130	The Council of Clermont brings an end to the practice of medicine by monks
1136	Pantokrator Hospital in Constantinople is founded by Byzantine Emperor John II
ca. 1140	Gerald of Toledo, Spain, translates hundreds of works of Aristotle, Avicenna, al-Razi, and Albacasis's writings on surgery
1150	University at Paris is founded
1158	University at Bologna is founded
1160	Abbess Hildegard of Bingen (woman practitioner) wrote *Liber simplicis medicine* (Simple book of medicine)
1167	University at Oxford founded
1181	University at Montpellier is founded
1189	Cesspit regulations become part of the zoning code in London
13th century	Five distinctions of the pulse: 1) motion of the arteries; 2) contraction of the artery; 3) diastolic and systolic duration and pressure; 4) strengthening or weakening of pulsation; 5) regularity or irregularity of the beat
	Italian physician inserts a cannula into the esophagus of a patient to serve as a feeding tube

	A few private homes have their own water supply piped into the home, and the inside water flow is controlled by a faucet
	A few communities add laws that require every citizen to visit a public bath weekly
	One out of every 200 Europeans suffer from leprosy
1200–1280	Albertus Magnus—first European to use the scientific method in his thinking and studies
1209	University at Cambridge is founded
1210–1280	Ibn an-Nafis develops the theory that the blood needs to go from the right ventricle to the lungs to acquire air, and then it enters into the left ventricle
1211	The town of Reggio, Italy, pays a public physician to help care for the poor, help with inquests, treat sufferers of the plague, and attend to injuries inflicted on prisoners
1222	University at Padua is founded
1225	There are 19,000 leprosaria in Europe
1226–1274	Thomas Aquinas introduces a new emphasis on Aristotle in the university curriculum
1250–1306	Lanfranc of Milan writes his *Chirugia Magna* (1296) consisting of his knowledge of anatomy, embryology, ulcers, fistulae, fractures, dislocated joints, and herbal medicines
1269–1320	Henri de Mondeville recommends closing wounds before pus has time to form
1298–1368	Guy de Chauliac (known as the father of surgery) writes a medical reference book about surgery in 1363
14th century	Arnold of Villanova focuses on improving elixirs such as alcohol. He comes to understand the existence of carbon monoxide.
	Italy begins to employ surgeons for autopsies

	Midwives are required to be licensed before practicing
	Quarantines are established to combat reoccurrences of bubonic plague
ca. 1300	False Gerber discovers vitriol (sulphuric acid) and describes how to make strong nitric acid
1307–1392	John of Arderne specializes in repair of fistula. Devises a method of cutting the lump out and making repairs
ca. 1315	Dissection becomes a compulsory part of the medical curriculum for the first time at the University of Bologna
1317	First pharmacy is created in Dubrovnik, Croatia
1322	In Paris, five women are put on trial for practicing medicine without a license. They are found guilty and excommunicated from the church.
1331	Bubonic plague outbreak in China
1347	First outbreak of bubonic plague (Black Death) in Europe
ca. 1393–1438	Margery Kempe writes *The Book of Margery Kempe,* depicting the relationship between practical medicine and spiritual care
1421	English physician Gilbert Kymer and some colleagues submit a petition to the British Parliament to ban women from working as doctors
1453	Fall of Constantinople
1486	Two Dominican friars, Jacob Springer and Heinrich Kramer, write *Malleus maleficarum* (The Witch Hammer) focusing on their pursuit of women who were practicing witchcraft

GLOSSARY

alchemy a medical chemical science and speculative philosophy aiming to achieve the transmutation of the base metals into gold, the discovery of a universal cure for disease, and the discovery of a means of indefinitely prolonging life

amulet a charm (as an ornament) often inscribed with a magic incantation or symbol to aid the wearer or protect against evil (as disease or witchcraft)

anatomy the act of separating the parts of the organism in order to ascertain their position, relations, structure, and function

anesthetic a substance that produces anesthesia; something that brings relief

apothecary one who prepares and sells drugs or compounds for medicinal purposes

aqueduct a conduit for water; especially one for carrying a large quantity of flowing water

artery any of the tubular branching muscular- and elastic-walled vessels that carry blood from the heart through the body

astrology the divination of the supposed influences of the stars and planets on human affairs and terrestrial events by their positions and aspects

bile either of two humors associated in old physiology with irascibility and melancholy; a yellow or greenish viscid alkaline fluid secreted by the liver and passed into the duodenum where it aids especially in the emulsification and absorption of fats

bloodletting phlebotomy—the letting of blood for transfusion, diagnosis, or experiment, and especially formerly in the treatment of disease

bubonic plague plague caused by a bacterium (*Yersinia pestis*) and characterized especially by the formation of buboes

caliphs a successor of Muhammad as temporal and spiritual head of Islam—used as a title

cataract a clouding of the lens of the eye or of the surrounding transparent membrane, obstructing the passage of light

cautery the act or effect of cauterizing; an agent (as a hot iron or caustic) used to burn, sear, or destroy tissue

cesspit a pit for the disposal of refuse (as sewage)

contagion a contagious disease; the transmission of a disease by direct or indirect contact

Crusades any of the military expeditions undertaken by Christian powers in the 11th, 12th, and 13th centuries to win the Holy Land from the Muslims

cupping an operation of drawing blood to the surface of the body by use of a glass vessel evacuated by heat

diagnosis the art or act of identifying a disease from its signs and symptoms

dissect to separate into pieces: expose the several parts of (as an animal) for scientific examination

dysentery a disease characterized by severe diarrhea with passage of mucus and blood and usually caused by infection

empirics charlatan; one who relies on practical experience

endemic belonging or native to a particular people or country

epidemic affecting or tending to affect a disproportionately large number of individuals within a population, community, or region at the same time

fast to abstain from food; to eat sparingly or abstain from some foods

folk medicine traditional medicine as practiced nonprofessionally, especially by people isolated from modern medical services and usually involving the use of plant-derived remedies on an empirical basis

Gaul a part of what is now France

guilds an association of people with similar interests or pursuits, especially a medieval association of merchants and craftsmen

gynecology a branch of medicine that deals with the diseases and routine physical care of the reproductive system of women

hernia a protrusion of an organ or part (as the intestine) through connective tissue or through a wall of the cavity (as of the abdomen) in which it is normally enclosed

hospital a charitable institution for the needy, aged, infirm, or young; an institution where the sick or injured are given medical or surgical care

humors a fluid or juice of an animal or plant; specifically one of the four fluids entering into the constitution of the body and determining by their relative proportions a person's health and temperament

incantation a use of spells or verbal charms spoken or sung as a part of a ritual of magic; *also:* a written or recited formula of words designed to produce a particular effect

infirmary a place where the infirm or sick are lodged for care and treatment

inflammation a local response to cellular injury that is marked by capillary dilatation, leukocytic infiltration, redness, heat, and pain and that serves as a mechanism initiating the elimination of noxious agents and of damaged tissue

kidney stone a calculus (as of calcium salts) in the kidney

leeches any of numerous carnivorous or bloodsucking, usually freshwater, annelid worms (genus *Hirundinea*) that have typically a flattened lanceolate segmented body with a sucker at each end

leprosy a chronic infectious disease caused by a mycobacterium (*Mycobacterium leprae*) affecting especially the skin and peripheral nerves and characterized by the formation of nodules or macules that enlarge and spread accompanied by loss of sensation with eventual paralysis, wasting of muscle, and production of deformities—called also Hansen's disease

liver a large very vascular glandular organ of vertebrates that secretes bile and causes important changes in many of the substances contained in the blood (as by converting sugars into glycogen which it stores up until required and by forming urea)

mandrake a Mediterranean herb *(Mandragora officinarum)* of the nightshade family with large ovate leaves, greenish-yellow or purple flowers, and a large usually forked root resembling a human in form and formerly credited with magical properties

measles an acute contagious disease that is caused by a morbilivirus (species *Measles virus*) and is marked especially by an eruption of distinct red circular spots—called also rubeola

miasma a vaporous exhalation formerly believed to cause disease; *also:* a heavy vaporous emanation or atmosphere (a ~ of tobacco smoke)

microbe microorganism, germ

midwife a person who assists women in childbirth

opium a bitter brownish addictive narcotic drug that consists of the dried latex obtained from immature seed capsules of the opium poppy

pandemic occurring over a wide geographic area and affecting an exceptionally high proportion of the population (~ malaria)

pharmacy the art, practice, or profession of preparing, preserving, compounding, and dispensing medical drugs

philosophy a discipline comprised of logic, aesthetics, ethics, metaphysics, and epistemology

phlegm the one of the four humors in early physiology that was considered to be cold and moist and to cause sluggishness

physician a person skilled in the art of healing; specifically one educated, clinically experienced, and licensed to practice medicine, as usually distinguished from surgery

pilgrimage a journey of a pilgrim, especially to a shrine or a sacred place

plague an epidemic disease causing a high rate of mortality

pores a minute opening especially in an animal or plant: specifically one by which matter passes through a membrane

psoriasis a chronic skin disease characterized by circumscribed red patches covered with white scales

pulse a regular expansion of an artery caused by the ejection of blood into the arterial system by contractions of the heart

pustule a small circumscribed elevation of the skin containing pus and having an inflamed base

quarantine a restraint upon the activities or communication of persons or the transport of goods designed to prevent the spread of disease or pests

septicemic characteristic of a systemic disease caused by pathogenic organisms or their toxins in the bloodstream; blood poisoning

smallpox an acute contagious febrile disease of humans that is caused by a poxvirus (species *Variola virus* of the genus *Orthopoxvirus*)

styptic tending to contract or bind: astringent; especially tending to check bleeding

surgery a branch of medicine concerned with diseases and conditions requiring or amenable to operative or manual procedures

topiary a plant clipped into fantastic shapes, often animals

transfusion the process of transfusing fluid (as blood) into a vein or artery

trephination the act or instance of perforating the skull with a medical instrument

tuberculosis a highly variable communicable disease of humans and some other vertebrates that is caused by the tubercle bacillus and, rarely in the United States, by a related mycobacterium *(Mycobacterium bovis)*, that affects especially the lungs but may spread to other areas (as the kidney or spinal column), and that is characterized by fever, cough, difficulty in breathing, formation of tubercles, caseation, pleural effusions, and fibrosis

tumor a swollen distended part; an abnormal benign or malignant new growth of tissue that possesses no physiological function and arises from controlled usually rapid cellular proliferation

urine waste material that is secreted by the kidney in vertebrates, is rich in end products of protein metabolism together with salts and pigments, and forms a clear amber and usually slightly acid fluid in mammals, but is semisolid in birds and reptiles

vaccine a preparation of killed microorganisms, living attenuated organisms, or living fully virulent organisms that is administered to produce or artificially increase immunity to a particular disease

vein any of the tubular branching vessels that carry blood from the capillaries toward the heart

FURTHER RESOURCES

ABOUT SCIENCE AND HISTORY

Diamond, Jared. *Guns, Germs, and Steel: The Fates of Human Societies.* New York: W. W. Norton and Company, 1999. Diamond places the development of human society in context, which is vital to understanding the development of medicine.

Hazen, Robert M., and James Trefil. *Science Matters: Achieving Scientific Literacy.* New York: Doubleday, 1991. A clear and readable overview of scientific principles and how they apply in today's world, including the world of medicine.

Internet History of Science Sourcebook. Available online. URL: http://www.fordham.edu/halsall/science/sciencsbook.html. Accessed July 9, 2008. A rich resource of links related to every era of science history, broken down by disciplines, and exploring philosophical and ethical issues relevant to science and science history.

Lindberg, David C. *The Beginnings of Western Science, Second Edition.* Chicago: University of Chicago Press, 2007. A helpful explanation of the beginning of science and scientific thought. Though the emphasis is on science in general, there is a chapter on Greek and Roman medicine as well as medicine in medieval times.

Roberts, J. M. *A Short History of the World.* Oxford: Oxford University Press, 1993. This helps place medical developments in context with world events.

Silver, Brian L. *The Ascent of Science.* New York: Oxford University Press, 1998. A sweeping overview of the history of science from the Renaissance to the present.

Spangenburg, Ray, and Diane Kit Moser. *The Birth of Science: Ancient Times to 1699.* New York: Facts On File, 2004. A highly readable book with key chapters on some of the most significant developments in medicine.

ABOUT THE HISTORY OF MEDICINE

Ackerknecht, Erwin H., M.D. *A Short History of Medicine, Revised Edition.* Baltimore, Md.: Johns Hopkins University, 1968. While there have been many new discoveries since Ackerknecht last updated this book, his contributions are still important as they help the modern researcher better understand when certain discoveries were made and how viewpoints have changed over time.

Arano, Luisa Cogliati. *The Medieval Health Handbook.* New York: George Braziller, 1976. Originally published in Italy as *Tacuinum sanitatis,* this book describes medieval cures of the day.

Clendening, Logan, ed. *Source Book of Medical History.* New York: Dover Publications, 1942. Clendening has collected excerpts from medical writings from as early as the time of the Egyptian Papyri, making this a very valuable reference work.

Davies, Gill, ed. *Timetables of Medicine.* New York: Black Dog & Leventhal, 2000. An easy-to-assess chart/time line of medicine with overviews of each period and sidebars on key people and developments in medicine.

Dawson, Ian. *The History of Medicine: Medicine in the Middle Ages.* New York: Enchanted Lion Books, 2005. A heavily illustrated short book to introduce young people to what medicine was like during medieval times. Dawson is British so there is additional detail about the development of medicine in the British Isles.

Dittrick Medical History Center at Case Western Reserve. Available online. URL: http://www.cwru.edu/artsci/dittrick/site2/. This site provides helpful links to medical museum Web sites. Accessed October 31, 2008.

Duffin, Jacalyn. *History of Medicine.* Toronto, Canada: University of Toronto Press, 1999. Though the book is written by only one author (a professor), each chapter focuses on the history of a single aspect of medicine, such as surgery or pharmacology. It is a helpful reference book.

Kennedy, Michael T., M.D., FACS. *A Brief History of Disease, Science, and Medicine.* Mission Viejo, Calif.: Asklepiad Press, 2004. Michael Kennedy was a vascular surgeon and now teaches first and second year medical students an introduction to clinical medicine course at the University of Southern California. The book started as a series of his lectures but he has woven the material together to offer a cohesive overview of medicine.

Loudon, Irvine, ed. *Western Medicine: An Illustrated History.* Oxford, England: Oxford University Press, 1997. A variety of experts contribute chapters to this book that covers medicine from Hippocrates through the 20th century.

Magner, Lois N. *A History of Medicine.* Boca Raton, Fla.: Taylor & Francis Group, 2005. An excellent overview of the world of medicine from paleopathology to microbiology.

Porter, Roy, ed. *The Cambridge Illustrated History of Medicine.* Cambridge, Mass.: Cambridge University Press, 2001. In essays written by experts in the field, this illustrated history traces the evolution of medicine from the contributions made by early Greek physicians through the Renaissance, Scientific Revolution, and 19th and 20th centuries up to current advances. Sidebars cover parallel social or political events and certain diseases.

Porter, Roy. *The Greatest Benefit to Mankind: A Medical History of Humanity.* New York: W.W. Norton Company, 1997. Over his lifetime, Porter wrote a great amount about the history of medicine, and this book is a valuable and readable detailed description of the history of medicine.

Rosen, George. *A History of Public Health, Expanded Edition.* Baltimore, Md.: Johns Hopkins University Press, 1993. While serious public health programs did not get underway until the 19th century, Rosen begins with some of the successes and failures of much earlier times.

Simmons, John Galbraith. *Doctors & Discoveries.* Boston: Houghton Mifflin Company, 2002. This book focuses on the personalities behind the discoveries and adds a human dimension to the history of medicine.

United States National Library, National Institutes of Health. Available online. URL: http://www.nlm.nih.gov/hmd/. Accessed July 10, 2008. A reliable resource for online information pertaining to the history of medicine.

OTHER RESOURCES

Annenberg Media Learner.org. Interactives. Available online. URL: http://www.learner.org/interactives/middleages/morhealt.html. Accessed October 31, 2008. Information on medieval medicine with links to other medieval sites.

Newman, Paul B. *Daily Life in the Middle Ages.* Jefferson, N.C.: McFarland & Company, 2001. This is a wonderfully thorough book about life in the middle ages, and it describes everything from what people ate to how they fought during medieval times.

Sacks, Oliver. *Migraine.* New York: Vintage Press, 1999. A helpful book about understanding migraine headaches that happens to refer to Hildegard's experiences.

INDEX

Note: Page numbers in *italic* refer to illustrations; *m* indicates a map; *t* indicates a table.

C

D

E

Q

R

S

W

Y

Z